NATUREGUIDE
TREES

SMITHSONIAN
NATURE GUIDE
TREES

Tony Russell

**LONDON, NEW YORK, MELBOURNE,
MUNICH, AND DELHI**

DORLING KINDERSLEY

Senior Editor Janet Mohun
Project Art Editor Anna Hall
Editor Miezan van Zyl
US Editors Daniel Letchworth, Rebecca Warren
Production Editor Rebekah Parsons-King
Production Controller Erika Pepe
Database David Roberts
New Photography Gary Ombler
New Illustrations Gill Tomblin
Picture Researcher Jo Walton
DK Picture Library Claire Bowers, Rose Horridge
Jacket Editor Manisha Majithia
Jacket Designers Laura Brim, Mark Cavanagh
Managing Editor Camilla Hallinan
Managing Art Editor Michelle Baxter
Publisher Sarah Larter
Art Director Philip Ormerod
Associate Publishing Director Liz Wheeler
Publishing Director Jonathan Metcalf

DK INDIA

Editorial Manager Rohan Sinha
Consultant Art Director Shefali Upadhyay
Deputy Managing Editor Alka Thakur Hazarika
Deputy Design Manager Mitun Banerjee
Senior Editor Anita Kakar
Senior Art Editors Arup Giri, Ivy Roy
Editors Rahul Ganguly, Priyanka Naib, Rupa Rao, Priyaneet Singh
Art Editors Vritti Bansal, Sanjay Chauhan, Arijit Ganguly, Rakesh Khundongbam, Pooja Pawwar, Vaibhav Rastogi
Assistant Editor Sudeshna Dasgupta
DTP Manager Balwant Singh
DTP Designers Rajesh Singh Adhikari, Dheeraj Arora, Neeraj Bhatia, Mohammad Usman, Anita Yadav
Production Manager Pankaj Sharma

SPECIAL PHOTOGRAPHY

The following locations in the UK were used for new photography:
Batsford Arboretum, Gloucestershire, UK
The Forestry Commission at Westonbirt, The National Arboretum, Gloucestershire, UK
The Royal Botanic Gardens, Kew, London, UK

CONSULTANT

Rusty Russell, Collections and Informatics,
United States National Herbarium, National Museum of Natural History,
Smithsonian Institution, USA.

First American Edition, 2012
Published in the United States by DK Publishing
375 Hudson Street, New York, New York, 10014

13 14   10 9 8 7 6 5 4 3
007 – 183145 – Apr/2012

Published in Great Britain by Dorling Kindersley Limited.

A catalog record for this book is available from the Library of Congress.

ISBN 978-0-7566-9039-7

DK books are available at special discounts when purchased in bulk for sales promotions, premiums, fund-raising, or educational use. For details, contact: DK Publishing Special Markets, 375 Hudson Street, New York, New York, 10014 or SpecialistSales@dk.com

Reproduced by Bright Arts, China, and MDP, UK
Printed and bound in China by Leo Paper Products

Discover more at **www.dk.com**

CONTENTS

HOW THE SPECIES PROFILES WORK

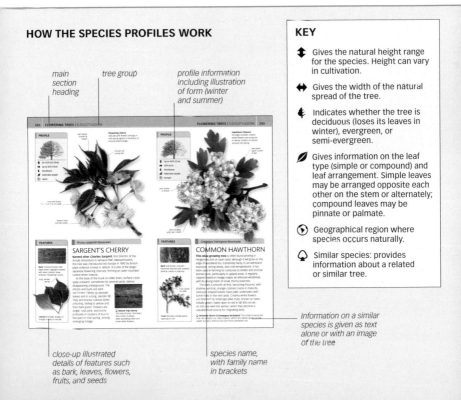

main section heading

tree group

profile information including illustration of form (winter and summer)

close-up illustrated details of features such as bark, leaves, flowers, fruits, and seeds

species name, with family name in brackets

Information on a similar species is given as text alone or with an image of the tree

KEY

Gives the natural height range for the species. Height can vary in cultivation.

Gives the width of the natural spread of the tree.

Indicates whether the tree is deciduous (loses its leaves in winter), evergreen, or semi-evergreen.

Gives information on the leaf type (simple or compound) and leaf arrangement. Simple leaves may be arranged opposite each other on the stem or alternately; compound leaves may be pinnate or palmate.

Geographical region where species occurs naturally.

Similar species: provides information about a related or similar tree.

WHAT ARE TREES?

Trees are distinguished from other plants by their upright woody stem, roots, and branches, and by the fact that they do not die back on a regular basis, but continue to grow year upon year. They are the largest plants on Earth.

ROLE OF TREES

Trees are essential to all life. They absorb vast amounts of carbon dioxide and other pollutants from the atmosphere, replacing them with oxygen. Forests help regulate excess water flow and can reduce the effects of flooding and soil erosion. Trees influence weather patterns by increasing humidity and generating rainfall. They help to green our cities and towns, and provide a habitat for wildlife.

Despite so many obvious values, large swaths of forest are being lost every year, often for commercial gain or farming. Now, more than 10 percent of the world's tree species are endangered. More than 8,750 species are threatened with extinction, and across the world nearly 100 acres of trees are felled every minute of every day.

Rainforest
There are thousands of different rainforest trees. Madagascar alone supports more than 2,000 different species, and there can be up to 200 different species within a single acre in the Amazon rainforest.

EVOLUTION

Trees evolved from early ferns in the Devonian. *Archaeopteris* was one of the first trees that made up most of the forests on Earth 300 million years ago. Fossil records show it had a woody trunk, up to 16 in (40 cm) in diameter, with branches and a large root system. It had a vital influence on the planet, removing massive amounts of carbon dioxide from the atmosphere and replacing it with oxygen. As a result, many new organisms evolved. As climate changed, tree species evolved and adapted to take advantage of the conditions. Modern trees are little changed since the Oligocene epoch (34–23 MYA) in the Paleogene.

Geological period	Million years ago	Event
TIMELINE		
Devonian	416–360 MYA	Spore trees emerge. *Archeopteris*, an early tree, dominates.
Carboniferous	360–299 MYA	*Lepidodendron*, the scale tree, flourishes in the warm, humid climate.
Permian	299–251 MYA	First primitive conifers (gymnosperms) emerge. They had protected seeds and an improved reproductive system.
Triassic	251–200 MYA	Conditions not favorable for tree growth but tree ferns, cycads, and conifers survive. Maidenhair tree emerges.
Jurassic	200–146 MYA	Cycads and conifers flourish.
Cretaceous	146–66 MYA	Flowering plants including most modern tree families evolve and begin to exert their dominance over conifers.
Paleogene	66–23 MYA	Flowering trees dominate and palms first appear. Grassland replaces large amount of forest; modern trees established.

TREE CLASSIFICATION

Classification is the grouping and naming of plants, or other living organisms. Scientists have tried to group trees for centuries, but not with great accuracy when compared with what scientists know today. The English oak (*Quercus robur*) and the holm or evergreen oak (*Quercus ilex*), have always been correctly regarded as being closely related to each other because of the fruit they produce—the acorn. However, the sweet chestnut (*Castanea sativa*) and the horse chestnut (*Aesculus hippocastanum*) were originally grouped together because of their similar fruits. Yet today they are classified as belonging to two quite separate families.

In the 18th century, the Swedish botanist Carl von Linné (Linnaeus) made the first attempt to link all plants by one specific feature. He classified them by the way they reproduced themselves: their sexual characteristics and the construction of their flowers and reproductive system.

He used two Latin words to name a species. The first word of the name is the generic name (genus)—for example, *Quercus*. This, when combined with the second or specific name (species)—for example, *robur*—becomes a unique identity for that species. Closely related species are all placed in the same genus, plural genera. Closely related genera are then grouped in families, which are in turn grouped in orders. Several orders together form a subclass, and subclasses are then combined under one class—for example, Angiospermae (flowering trees).

In recent years, a system of classification, based on genetic similarities, has been introduced. As a result, some family relationships have changed.

Major divisions

In this book, chapters are divided into spore trees and seed trees. Seed trees are then are further subdivided into the major divisions: ginkgos, conifers, and flowering trees. Flowering trees are further subdivided into primitive angiosperms, monocotyledons, and eudicotyledons.

```
              SEED              SPORE
              TREES             TREES
                │                 │
                ○
         ┌──────┴──────┐
    GINKGOS    CONIFERS   FLOWERING
                            TREES
                              │
                              ○
         ┌──────────┬──────────┐
  PRIMITIVE    MONOCOTYLEDONS  EUDICOTYLEDONS
 ANGIOSPERMS
```

CLASS	ANGIOSPERMAE
SUBCLASS	EUDICOTYLEDONEAE
ORDER	SAPINDALES
FAMILY	SAPINDACEAE
GENUS	ACER
SPECIES	ACER PALMATUM

Linnaean system
In this system, all species are classified by genus, family, and order. The example here shows how Japanese maple (*Acer palmatum*) fits into the order Sapindales.

ACER PALMATUM

CULTIVARS AND HYBRIDS

A hybrid occurs when two species cross-fertilize and produce offspring that display characteristics different to both parents. This can happen naturally in the wild, or be artificially induced. Hybrids are represented by "×" between the genus and species name. With cultivars, a difference in a species' characteristics can only be maintained through horticultural cultivation.

***Quercus rubra* 'Aurea'**
This cultivar of the red oak (*Quercus rubra*) has yellow leaves in contrast to the deep green leaves of the pure species.

HOW TREES WORK

There are thousands of different tree species but they all share one feature that distinguishes them from other plants: they produce a woody stem (a trunk) that supports a framework of branches and continues to grow year after year.

TREE STRUCTURE

Trees are made up of many different parts, and each part has its own unique function without which the tree could not survive. Some parts, such as the trunk, branches, and leaves, are clearly visible. Others parts are less obvious, for example, the circulatory system, roots, and, in some species, the reproductive organs.

canopy

trunk

Parts of a tree
The main parts of a tree above ground are the trunk and the canopy—the framework of branches bearing leaves, flowers, and seeds.

Seeds
The next generation of trees, seeds contain the basic ingredients for creating new trees virtually identical to their parents. Seeds come in various forms and may be contained in fruits, such as berries, nuts, or cones.

Flowers
The flowers contain the tree's reproductive organs. Male and female parts may occur within the same flower, or the tree may have separate male and female flowers either on the same or different tree.

Leaves
These produce food used by the tree to provide energy for living and growing. This is achieved by photosynthesis (opposite). Leaves are also the site of transpiration (water loss by evaporation).

Bark
A waterproof layer that surrounds the trunk and branches and protects the interior of the tree. It has millions of tiny pores (lenticels) that allow oxygen to pass into the living cells inside.

Roots
Tree roots provide anchorage and are the means by which the tree obtains water, minerals, and nutrients from the soil. Roots also store food, in the form of starch, for later use.

MAKING FOOD

Food, in the form of energy-rich sugars, is supplied through a process called photosynthesis that takes place in the leaves. Nutrients are also transported from the soil via the roots. The sugar (sucrose) from photosynthesis provides energy and can be turned into starch and cellulose, which form the tree's cells. Oxygen is a by-product of the process and is released back into the atmosphere.

sunlight activates chlorophyll inside leaf cell

chlorophyll in leaf cell converts water and carbon dioxide into sugar

oxygen released though stomata (pores) on underside of leaf

sugars carried back into the tree for growth

Photosynthesis
Chlorophyll, the green pigment in leaves, is activated by light energy from the sun. Tiny pores in the leaf's surface absorb carbon dioxide from the atmosphere. This reacts with water to produce sugar.

carbon dioxide from air enters leaf

water and minerals transported from roots to leaf cells

TREE GROWTH AND CIRCULATION

Trees grow from the tips of the previous year's growth. At the tip of each branch are growing cells. As these multiply, the branch grows longer and the tree taller and broader. Cells also divide beneath the bark, making the trunk and branches thicker.

Under the bark is a spongy layer called phloem. This carries the food made in the leaves (above) to the rest of the tree. Under the phloem is a layer of cells called the cambium, which makes phloem cells on the outside and xylem cells on the inside. Xylem transports water and minerals from the roots to the leaves.

Insect pollination
Pollen is transferred from the male reproductive structures (stamens) as the insect retrieves sweet nectar from the base of the flower.

REPRODUCTION

Tree flowers last just long enough to be pollinated by insects, birds, or wind. Once pollinated, and the flower's ovules (female parts) are fertilized, a seed develops from the female flower or part of flower. Depending on tree species, the seed matures within a fruit or cone. Once the seed is mature, it may fall to the ground, or be dispersed by wind or water. Some seeds are eaten by birds or mammals and excreted away from the tree. As long as conditions are suitable, the seed will eventually germinate and a seedling will develop—producing the next tree generation.

Tree rings
The tree rings seen in a cut trunk represent a new layer of xylem, which is deposited during each year's growing season. These bands can be used to tell a tree's age.

TREE HABITATS

Across the world, different tree species grow naturally in various locations that suit their specific requirements. These locations are known as their habitat—their natural home. What makes a location exactly right for a species will depend on many factors.

ADAPTING TO THE ENVIRONMENT

Over millions of years, each tree species has evolved and adapted to the climate and conditions that surround it. The amount of rainfall, the temperatures they have to endure, and the amount of sunshine they receive have all influenced both the behavioral patterns of trees and their natural distribution across the planet. Other influential factors include soil type, proximity to the oceans, and exposure to predation, disease, and fire. Trees show some remarkable adaptations that allow them to survive in a wide variety of conditions throughout the world. For example, mangrove trees (p.328), which grow close to, or actually in the ocean, have developed an internal desalination process that allows them to extract fresh water from the sea.

FOREST ZONES

Global forest zones
Broadly speaking, the world can be divided into four main forest zones. These zones are interspersed by areas that contain comparatively few trees, such as deserts and high mountain ranges.

KEY

- Coniferous forest
- Temperate broadleaf forest
- Tropical broadleaf forest
- Tropical rainforest
- Barren landscapes/deserts
- Ice caps
- Prairie and savanna

Coniferous forest covers large swaths across the northern hemisphere and some mountain ranges. Conifer trees are well-adapted to harsh winters.

Tropical broadleaf forest includes all tropical forest types, except for rainforest. Trees that fit in this category may also thrive in other zones.

Temperate broadleaf forest comprises a mix of evergreen and deciduous species, which tolerate the cool conditions that result from seasonal changes.

Tropical rainforest Is home to plants that thrive only in areas that have very high humidity, such as tropical South America.

TROPICAL

Tropical trees are located on either side of the equator, in three main regions: Amazonia in South America, Central Africa, and Southeast Asia. Here, rainfall, temperature, and hours of sunlight are relatively constant throughout the year and tree growth is rapid. Consequently, tropical trees reach maturity, flower, set seed, and die much more quickly than trees in other regions. A variety of characteristics help tropical trees grow in this habitat. For example, many trees, such as the weeping fig (p.314), have leathery leaves with pointed "drip tips," which allow them to shed water during tropical rainstorms. As there is little need for protection against frost or water loss, thin bark, as on the rubber tree (p.329), is another common feature. To gain maximum sunlight, while also shading out competing trees, some tropical trees, such as the Brazil nut (p.322), develop a tall trunk with few low branches and a large crown of leaves.

Dense rainforest
There is no dry season in areas where tropical rainforests have formed, which means that plant growth is continuous throughout the year. Rainforests are home to a very wide variety of plants.

Weeping fig leaves
The leaves of the weeping fig (*Ficus benjamina*) are ovate to elliptic, with a pronounced "drip tip" at the end. During the heavy rainfall that is common in the tropics, this tip allows excess water to run off.

›› TEMPERATE

Temperate trees are found between latitudes 40–50 degrees north and south of the equator. Here there are well defined seasons but few extremes in either rainfall patterns or temperature. A temperate climate is suitable for tree growth for approximately six months of the year, when average temperatures are over 50°F (10°C). Although suitable for both deciduous and evergreen trees, including conifers, it is deciduous broad-leaved trees, such as oak (p.165) that predominate in temperate regions. Many trees, such as birch (*Betula* sp., p.173) and hornbeam (p.184), take advantage of the windy conditions that exist within temperate regions and are wind pollinated.

Fall colors
In the fall, the leaves of deciduous trees undergo a chemical process as they are about to be shed, producing a wide array of color.

Winter forest
Deciduous broad-leaved trees discard their leaves in fall, growing new ones in the following spring when temperatures and light levels become suitable for growth.

DESERT

The difficulties facing trees in the desert all relate to a lack of water. Hot sun, desiccating winds, and low rainfall make it hard for roots to supply enough water to compensate for that lost through leaves. Desert trees have adapted to this harsh habitat in several ways. Some, such as the dragon tree (p.120), have leaves that reduce the transpiration effect. Other trees develop long roots, which allow them to reach moisture near the water table. The baobab (p.319) can store dozens of gallons of water in its broad trunk, which significantly varies its trunk circumference between wet and dry seasons.

Joshua tree
The Joshua tree has spikelike, evergreen leaves that occur only at the tips of the branches, reducing the loss of moisture through the leaves.

MOUNTAINS

In many ways mountains have the same climate as that of the subarctic, although this depends on elevation and latitude. They usually have short summers, cold winters, and a mean temperature that rarely rises above 50°F (10°C). Wind speeds also tend to be much greater at high altitudes and the combination of drying winds and shallow soils, which can be frozen for long periods of time, ensure that only the hardiest of trees will survive. The higher the elevation, the slower trees grow and the more stunted they become. The point beyond which no tree will survive is known as the tree line. The narrow needlelike leaves of many conifers are far better able to withstand freezing temperatures than broad leaves. Their narrow conical form, such as on Serbian spruce (p.83), also offers reduced wind resistance and the ability to shed heavy falls of snow.

Mountain pine
Mountains tend to be covered with conifers, many from the Pinaceae family—spruce, pine, fir, and hemlock—as they are better adapted to cold conditions.

»

» COASTAL TREES

Exposure to ocean storms and strong salt-laden winds can make it difficult for coastal trees to survive. Salt contact with foliage may cause leaf-burn, branch die-back, and defoliation. This reduces the tree's ability to photosynthesize and produce food, causing stress and possible death. Salt uptake from moisture extracted by tree roots from coastal soils can also poison a tree. It takes only 0.5% of a tree's living tissue to contain salt before it reaches toxic levels. Some trees, such as Monterey cypress (p.51), which grows on the rocky coastline of the Monterey Peninsula in California, are able to absorb higher concentrations of salt without harm.

Coastal cypress
Trees that grow along the coast must be able to withstand harsh conditions caused by strong winds and salt. The Monterey cypress (*Cupressus macrocarpa*) has a higher threshold against salt toxicity than many other trees.

Mangrove roots
Mangroves (*Rhizophora* spp.) produce stilt or prop roots that rise above the water to absorb oxygen. The roots also screen out salt.

Urban
Tree-lined streets and urban parks improve air quality, reduce heat, and help to combat noise and light pollution.

URBAN TREES

Trees bring many benefits to the urban environment. They reduce air pollution by trapping particle pollutants such as dust, ash, and smoke. They also absorb carbon dioxide and other dangerous gasses, replacing them with clean oxygen. Urban trees help reduce localized flooding by capturing rainwater and increasing soil permeability. They cool the "heat island" effect of cities by absorbing heat and providing shade. Trees reduce wind speed around high-rise buildings, increase humidity in dry city climates, and reduce glare on sunny days. Urban trees also reduce noise and light pollution. Trees such as *Ginkgo biloba* and London plane (*Platanus* x *acerifolia*) are perfect for urban planting because they are extremely tolerant of localized air pollution.

HEDGEROW

Hedgerows are live barriers consisting of trees (and shrubs) that are used to contain livestock, act as boundary markers, and provide shelter. The best trees for hedgerows are those species that will naturally develop a thick, impenetrable framework of branches, ideally protected by an armory of thorns or spines. They should be tough, able to withstand exposure, browsing by livestock, and respond well to regular pruning. The most popular trees for planting in hedges are hawthorn, holly, and beech. Hedgerows are valuable to wildlife and provide an ideal nesting habitat for birds.

Living wall
The trees growing within hedgerows differ from region to region and reflect the local conditions and the purpose of the hedge.

WETLAND

The banks of rivers and streams can provide the ideal habitat for many trees. Rivers provide a natural gap in the canopy, which means trees that grow alongside receive plenty of light. Wetland areas are also rich and fertile—flooding and constant fluctuations in water levels ensure regular deposits of minerals and nutrients. However, during prolonged flooding the soil will become waterlogged, preventing oxygen from getting to roots. Some trees, such as the swamp cypress (p.66), have adapted ways to overcome this. They produce special hollow breathing organs called pneumatophores, or "knees," which rise above the water. These knees have lenticels (breathing pores) in the bark that absorb oxygen and then pipe it back down to the roots beneath the water.

Swamp cypress
Taxodium distichum can withstand waterlogged conditions for many months. In the Florida Everglades, they form extensive forests where few other trees would survive.

IDENTIFYING TREES

Despite the large numbers of different species and varieties, trees offer several clues to their identification. By working through these step-by-step, it should be possible to reach a positive identification.

BARK

The great advantage of bark as an identification feature is that it is present all year round. So even in the depths of winter it is possible to identify some trees just by their bark. Color, flaking, banding, the development of ridges, and the appearance of lenticels (pores) on the bark can all aid identification. However, bark often changes with age, and this is an important factor to take into consideration when identifying a tree.

HORIZONTAL PEELING

VERTICAL PEELING

IRREGULAR PLATES

RIDGED AND FISSURED

SMOOTH IN EARLY YEARS

ROUGH FLAKING

Types of bark
The textures and color of bark are useful aids to identification. Throughout this book, these terms are used to descibe the bark of different species.

LEAF ARRANGEMENT

After assessing whether leaves are deciduous or evergreen, the way the leaves are arranged on the shoot is a helpful aid to identification. Some trees have large leaves that are divided into smaller leaves, or leaflets. These are compound leaves, and the arrangement of the leaflets provides another clue to identification. Pinnate leaves have leaflets on either side of a central stalk. Palmate leaves have leaflets arranged like fingers on a hand. Leaves that are not divided into leaflets are called simple. Simple leaves may be arranged opposite each other on the stalk or may be alternate.

OPPOSITE

ALTERNATE

PINNATE

PALMATE

Simple leaves
Simple leaves do not have leaflets and are arranged in one of two ways: opposite, with leaves paired on either side of the stalk, or alternate, with the leaves in a staggered pattern.

Compound leaves
Compound leaves are divided into leaflets, which may branch from the tip of the stalk (palmate) or may be arranged symmetrically along the stalk's length (pinnate).

LEAF SHAPE

Collectively, trees display scores of different leaf shapes and a wide range of sizes, from less than an inch to more than a yard. Broadleaf trees display the greatest variation in shape, but many are either elliptic or obovate. Conifer leaves tend to fall into just a few main categories: scalelike; linear; or needlelike and tapering to a pointed, but not always sharp, end.

NEEDLELIKE   **LINEAR**   **SCALELIKE**

OBLANCEOLATE   **OBLONG**   **LANCEOLATE**

OBOVATE   **ELLIPTIC**   **CORDATE (HEART-SHAPED)**   **ROUNDED**

Common leaf shapes
A selection of some of the most common leaf shapes is shown here. Throughout this book, the leaf shape of each tree species is given.

LEAF MARGINS AND TEXTURE

The margins of both compound and simple leaves may be smooth (known as entire) or may have indentations. The leaf margin indentations may be wavy, serrated or toothed (like the teeth of a saw), or lobed. One of the most recognizable leaf margins is that of the common oak (*Quercus robur*), which has regular "bites" along the margin interspersed with rounded lobes. Some leaves are also recognizable by their texture. For example, leaves of the English elm (*Ulmus procera*) have a rough texture, whereas those of the katsura tree (*Cercidiphyllum japonicum*) are velvetlike.

Types of leaf margin
Leaf margins may be smooth (entire), wavy, or indented. There are many variations of indentations, some of which are shown here.

ENTIRE   **WAVY**   **TOOTHED**   **LOBED**

»

›› FLOWERS

The flowers of many trees have a distinctive appearance and provide valuable clues to identification. In temperate regions the majority of trees flower in the spring, whereas in tropical regions flowers may be produced all year round. The purpose of flowers is to aid reproduction, and flowers tend to be colorful and fragrant in order to attract pollinators such as insects and birds. Some tree species have separate male and female flowers on the same tree; these are known as monoecious. Species in which the male and female flowers are on different trees are known as dioecious.

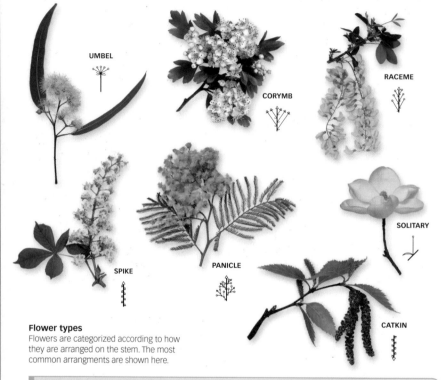

Flower types
Flowers are categorized according to how they are arranged on the stem. The most common arrangments are shown here.

PARTS OF A FLOWER

Most flowers consist of petals, sepals, and reproductive structures. The female parts are the stigma, style, and ovary; the male parts are the anther and filament (together forming the stamen). In most flowers, the sepals and petals are distinct, in primitive angiosperms such as magnolia, they are not.

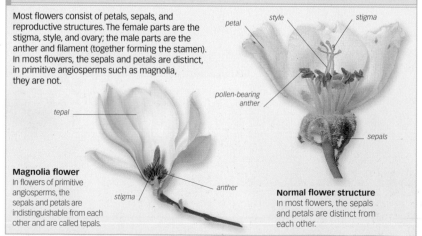

Magnolia flower
In flowers of primitive angiosperms, the sepals and petals are indistinguishable from each other and are called tepals.

Normal flower structure
In most flowers, the sepals and petals are distinct from each other.

SEEDS AND FRUIT

Nearly all trees begin life as seeds. These seeds are contained within a receptacle, or fruit, of which there are several types, including drupes, berries, pods, and nuts. Observing how a tree carries its seeds can be a useful aid to identification.

Most conifers encase their seeds in a hard, woody conical structure—a cone. Beneath the scales of each cone there may be hundreds of seeds, each attached to a papery wing to aid wind dispersal. Members of the *Acer* (maple) genus also produce seeds that are attached to papery wings, these are called samara.

In some instances, tree seeds are encased in a fruit of soft flesh. Where this flesh surrounds a single hard shell, which in turn encases a single seed, such as in peach, plum, olive, or cherry, the fruit is known as a drupe. When the flesh surrounds multiple seeds the fruit is known as a berry or a pome. Dry fruits that split on maturity to release the seeds inside are commonly called a pod or capsule. Dry fruits that do not automatically split include those fruits commonly known as nuts. Examples of these include hazel, hickory, chestnut, and acorn.

Types of seeds and fruit
The form, shape, size, and color of fruit and seeds can be an aid to tree identification.

berry

samara

PAPERBARK MAPLE

pod

JUDAS TREE

nut

PEDUNCULATE OAK

MIDLAND HAWTHORN

PLUM

drupe—flesh encases a single seed

cone

ATLAS CEDAR

TREE HABIT

The shape of a tree, known as its habit, can be a good clue to identification as some trees have a habit that is easily recognizable, such as the narrow, column-like habit of the Italian cypress (*Cupressus sempervirens*) and the weeping habit of the willow (*Salix babylonica*). A tree's size may also aid identification. However, both habit and size can vary according to factors such as a tree's age, habitat, and climate.

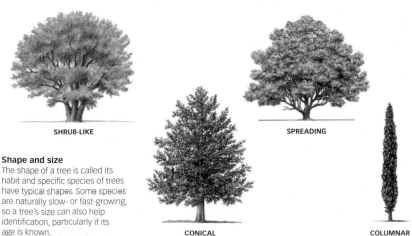

SHRUB-LIKE

SPREADING

Shape and size
The shape of a tree is called its habit and specific species of trees have typical shapes. Some species are naturally slow- or fast-growing, so a tree's size can also help identification, particularly if its age is known.

CONICAL

COLUMNAR

SPECIES GUIDE

SPORE TREES

The most primitive trees—the tree ferns—do not grow from seeds, but from single cells called spores. This type of life cycle has more in common with lowly mosses than other kinds of trees.

The very first trees—from around 300 million years ago—were spore trees, such as *Archeopteris* (p.6). Their trunks lifted foliage clear of the prehistoric swamps, giving them more access to sunlight. More light meant more energy to make more food. These primeval forests had many trees unlike anything around today. The only surviving spore-producing trees are tree ferns. Tree ferns have trunks covered in a mantle of aerial roots and a core strengthened by fiber, not wood. They predominate in certain habitats, such as equatorial alpine meadows—but otherwise live in forests dominated by seed trees.

Like all ferns, tree ferns scatter spores from ruptured capsules on the underside of leaves. Spores germinate on wet ground, each growing into a tiny heart-shaped sexual stage called a prothallus. Sperm cells from one prothallus swim to fertilize the egg in another and sprout an embryonic fern plant.

DIVISION	PTERIDOPHYTA
CLASS	FILICOPSIDA
ORDER	CYATHEALES
FAMILIES	8
SPECIES	c.680 (c.600 SPORE TREES)

Over many years of growth, a tree fern may come to tower 65 ft (20 m) over neighboring plants, but the nature of its life cycle means that it will always demand moisture on the ground to reproduce. This kind of spore-dependent reproduction occurs in other plants—such as mosses and liverworts—but none of these has evolved the tree habit.

All ferns belong to the plant division Pteridophyta. Less than 20 percent of fern species belong to the one tree fern order, which is called Cyatheales. The most primitive Cyatheales are trunkless ferns with creeping stems, but those of the families Dicksoniaceae and Cyatheaceae are tree fern "giants," with tough leathery leaves.

PROFILE

- 22 ft (7 m)
- Up to 12 ft (4 m)
- Evergreen
- Compound
- S.E. Australia, Tasmania

Large, young fronds
This tree fern has large fronds, up to 12 ft (4 m) long, which grow from the top of the trunk.

unfurling fern frond

FEATURES

Bark Chestnut-brown; fibrous, soft, with black, woody core

Leaves Slightly shiny, rich green on upper surface, matte, pale green on underside

Dicksonia antarctica (Dicksoniaceae)

SOFT TREE FERN

Native to southeastern Australia and Tasmania, this hardy tree fern has been popular for planting in the warm, wet regions of both Europe and North America since its introduction and cultivation in Europe in 1880. This species is not a tree at all but a true fern; however, its phenomenal popularity lies in the fact that it can reach treelike proportions.

The trunk is made of the fibrous remains of old roots. Beneath the trunk's surface is a bone-like core. New roots grow down the trunk each year to enable the fern to produce a new set of fronds. Each frond is like a typical fern leaf, but larger in size. The underside of fronds have light brown, rounded spores dispersed at regular intervals. New fronds unfurl from the top of the trunk in spring. In winter, the plant resembles an upright log because in colder climates the fronds die completely.

♤ **Rough tree fern (*Cyathea australis*)** This related tree fern is more cold-hardy, and sheds its fronds each year, even in warm climates, unlike the soft tree fern.

SEED TREES

The majority of living trees start their lives bound inside a capsule called a seed. Seeds provide a means of securing the welfare of embryonic trees until they are ready to sprout roots and shoots.

Seed trees are the largest group of trees and include the divisions cycads, ginkgos, gnetophytes, conifers, and the biggest group of all, flowering trees (also called angiosperms). Seed trees combine two key adaptive features: woody trunks and seeds. Wood is built from millions of microscopic hollow struts—called xylem vessels—that carry water and minerals from the roots in the ground to the leaves. The xylem vessels form rings around the core of the trunk, increasing its girth and helping it support more foliage.

SUPERDIVISION	SPERMATOPHYTA
DIVISIONS	CYCADOPHYTA
	GINKGOPHYTA
	GNETOPHYTA
	PINOPHYTA
	MAGNOLIOPHYTA

The seed is a survival capsule containing a plant embryo (a young plant at an early stage of development). Whereas exposed tree fern (p.23) embryos are vulnerable, embryos inside seeds are protected and are even nourished by a food store—up to the point when conditions are right for them to germinate. Seed trees can reproduce in drier habitats too, as their fertilization happens inside reproductive shoots of the canopy—not on wet ground. Each egg forms inside a female capsule called an ovule. Male cells are delivered in pollen grains from male shoots—carried either on wind, by water, or by animal, such as an insect. A "captured" pollen grain penetrates the ovule with a pollen tube to convey male cells to the egg. The fertilized egg then grows into an embryo and the ovule is now called a seed.

Most seed trees produce their seeds encased in fruit that develops from a flower: these belong to the class Angiospermae within the division Magnoliophyta. Others, such as Ginkgophyta (ginkgos) and Pinophyta (conifers)—collectively called gymnosperms—produce "naked" seeds.

GINKGOS

Most non-flowering seed trees alive today are conifers, but a few are remnants of groups that flourished even before the development of conifers. One of these groups is the ginkgos, the sole living representative of which is the maidenhair tree.

In contrast to the needlelike leaves of many conifers (p.27), the maidenhair tree has flat leathery leaves that are shed in harsh winters. Ginkgos have extraordinary reproductive behavior quite unlike any other plant alive today. They

DIVISION	GINKGOPHYTA
CLASS	1
ORDERS	1
FAMILIES	1
SPECIES	1

produce seed without cones, flowers, or fruits. Each male sex cell (sperm) produced by a pollen grain moves with a tuft of tails—a remnant of primitive fernlike ancestry. Ginkgos are unisexual: each individual carries either male or female structures, but never both. Female ginkgos produce their small egg-containing ovules in exposed pairs. Pollen grains reach them by wind and are "captured" by drops of fluid secreted by the ovules, which also serve to encourage the grains to grow pollen tubes: sperm then swims through the tube to fertilize the egg. Because the ovules are "naked" on the female shoots, ginkgos do not produce true fruits. Instead, their fleshy seeds smell of vomit and rotting flesh—something that humans find nauseating, but might have lured in a long-extinct carnivore to aid dispersal.

Primitive leaf
Unlike broadleaved trees, leaves of the ginkgo have veins that spread out like a fan, instead of forming a network.

*leaves often
deeply notched*

*leaf up to 3 in
(8 cm) wide*

↕ Up to 130 ft (40 m)

↔ Up to 70 ft (20 m)

🍂 Deciduous

🌿 Fan-shaped

🔄 S.E. China

*leaves spirally arranged
on new shoots*

*shoots form
along thin,
fawn-gray twigs*

*veins radiate
from base
of leaf*

Ginkgo leaves
The notched, leathery leaves of
the maidenhair tree are borne
on long, slender stalks and have
a distinctive fan shape.

FEATURES

Bark Pale gray-brown;
vertically fissured when mature

*seed up
to 1½ in
(4 cm) long*

*edible
kernel*

Seeds Fleshy, yellow-green
seed husk contains kernel

Leaves
Golden yellow
coloration
in fall

Ginkgo biloba (Ginkgoaceae)

MAIDENHAIR TREE

Fossil records indicate that this primitive tree has
existed on Earth since the Jurassic period, about 200
million years ago. It is no longer found in the wild, but the
presence of ancient cultivated trees in southeast China
suggests that the last wild population grew in this area. The
tree was introduced in the West in 1730. It produces leaves
resembling those of the maidenhair fern (*Adiantum*), hence
its common name.

Several leaves emerge,
spirally arranged, from each
shoot. The yellow-green male
and female reproductive organs
appear on separate trees.
Female trees are seldom grown
as the flesh of the seed emits a
rancid odor. The leaves have
medicinal properties and are
used in the treatment of certain
circulatory ailments.

Ornamental tree
The upright, broadly conical
Ginkgo is planted as an
ornamental tree throughout
the northern hemisphere.

CONIFERS

Conifers are large woody, sometimes massive, seed trees that have their reproductive parts arranged in shoots called strobili, or cones, instead of flowers. Most are evergreen and have small, but very tough, needlelike leaves that help them survive harsh cold winters or prolonged drought.

These trees are particularly resilient and many conifers are adapted for unforgiving conditions. Needle leaves, laden with resin and liberally coated in transparent wax, dry out very slowly—so little water is required to keep them alive. This means conifers can stay green when soil is frozen, and the same needle-type leaves helps tropical species survive where land is parched.

DIVISION	PINOPHYTA
CLASS	1
ORDERS	1
FAMILIES	7
SPECIES	630 (c. 550 TREES)

Conifers produce seeds in cones. Male cones produce pollen, female ones produce ovules—quite often on the same tree. Lacking the means to attract insects, they rely instead on the wind to carry their pollen. Wind pollination is a hit-and-miss affair, but conifers make up for it by producing their pollen in vast quantities. Ovules, and therefore seeds, arise on scales on female cones. In some conifers, such as junipers and yews, the cones are fleshy and edible to attract animals who disperse seed in their feces. In others, such as pines, hard woody cones may take years to open to release their seeds.

Lodgepole pine
Resinous leaves and cones help this hardy conifer survive in the coldest conditions, such as Tongass National Forest in Alaska.

PROFILE

scale tip
curved outward

Monkey puzzle cones
The male cone remains on the tree for many months. Male and female cones are normally carried on separate trees.

glossy dark green leaves, arranged spirally

male cone 3–6 in (7–15 cm) long

PROFILE

- ⬍ Up to 100 ft (30 m)
- ⬌ Up to 50 ft (15 m)
- 🕯 Evergreen
- 🌿 Triangular
- ◉ S. Chile, S.W. Argentina

FEATURES

developing female cone cluster

leaf 1¼–1½ in (3–4 cm) long, ⅜–1 in (0.8–2.5 cm) wide

Female cones
Green, ripening to yellow-brown

Seed cones Brown; breaking to release up to 200 seeds

Araucaria araucana (Araucariaceae)

MONKEY PUZZLE TREE

Also known as Chile pine, this distinctive tree has existed for millions of years. Fossil remains indicate it was abundant in the southern hemisphere during the Mesozoic era. Today, its natural range is exclusive to southern Chile and Argentina. It was introduced into cultivation in the 18th century using seeds collected by Archibald Menzies, the Scottish surgeon aboard the British ship HMS *Discovery*.

Mature trees have cylindrical, gradually tapering trunks, with a light gray bark that is slightly wrinkled. The triangular, stiff leaves terminate in sharp points. Spherical female cones are borne on the top side of branches, high in the crown. They mature over two years, growing up to 6 in (15 cm) wide, and break apart to release the seeds.

Distinctive shape
The trunk of the monkey puzzle tree has whorls of horizontal branches emerging at regular intervals.

Dense foliage
Leaves of the Norfolk Island pine are densely arranged along light brown branches, which are borne in widely spaced whorls.

stiff, horizontal to slightly ascending branches

leaf 1.2 cm (½ in) long

PROFILE

- 160–230 ft (50–70 m)
- Up to 80 ft (25 m)
- Evergreen
- Needlelike
- Norfolk Island (east of Australia)

FEATURES

Female cones Green, turning dull brown in 18 months

Araucaria heterophylla (Araucariaceae)

NORFOLK ISLAND PINE

This ancient tree is endemic to Norfolk Island, a small island to the east of Australia. Particularly attractive in its juvenile stage, this tree is grown in tropical and subtropical gardens, and is widely grown as a houseplant elsewhere. Prone to frost damage, this tree is rarely successful when grown outdoors at latitudes above 45° North.

The scientific name *heterophylla* means "different leaves," referring to the leaf variation between juvenile and adult trees. When young, the tree has light green, drooping, soft, awl-shaped leaves. With maturity, it develops a symmetrical, triangular outline, while the leaves become hard and dark green. Male pollen cones grow to 1½ in (4 cm) long. They are yellowish brown, sometimes reddish in color and borne in pendulous clusters. Female cones are squat, rounded, 4–5 in (10–12 cm) long and wide.

◊ **Kauri pine (*Agathis australis*)** This tree is easily distinguished from the Norfolk Island pine by its thick, bladelike leaves, and large spreading branches.

Leaves and cones
The leaves of the Wollemi pine are light green at first, turning dark green with age. The male cones are slender and ovoid.

leaf up to 1½ in (4 cm) long

male cone, up to 3 in (7.5 cm) long

‡ 70–130 ft (20–40 m)
↔ Up to 30 ft (10 m)
❦ Evergreen
⬦ Needlelike
◉ S.E. Australia

FEATURES

Bark Develops chocolate brown, bubbly texture with age

Female cones Small, green; take about two years to mature

Wollemia nobilis (Araucariaceae)

WOLLEMI PINE

Thought to have been extinct for at least two million years, this upright conifer was discovered in 1994, growing in a remote canyon in the Blue Mountains of Australia. It is related to the *Araucaria* and *Agathis* genera. Fossilized remains dating back approximately 90 million years indicate that the Wollemi pine was prevalent in the Cretaceous period. Its discovery has been described as the botanical equivalent to finding a living dinosaur. There are fewer than 100 mature trees in the wild, but thousands have been reproduced from them and are being cultivated around the world.

The bark is reddish brown, and peels in thin scales when young. Young leaves are soft and frond-like, while mature ones are linear, pointed, and leathery, and are arranged in four rows. The Wollemi pine produces both male and female cones on the same tree. The cones are initially green and mature to bronze. In winter, the growing tips of the buds are protected by a white, waxy coating and the leaves take on a bronze hue.

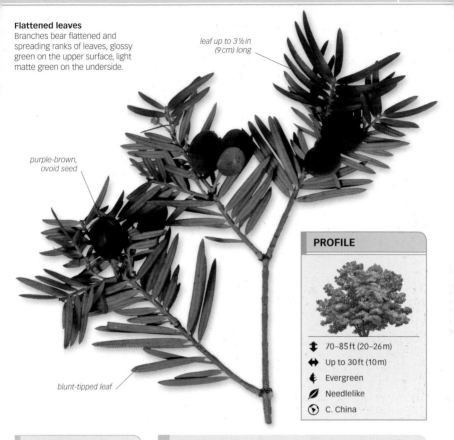

Flattened leaves
Branches bear flattened and spreading ranks of leaves, glossy green on the upper surface, light matte green on the underside.

leaf up to 3½ in (9 cm) long

purple-brown, ovoid seed

blunt-tipped leaf

PROFILE

↕ 70–85 ft (20–26 m)

↔ Up to 30 ft (10 m)

❦ Evergreen

🌿 Needlelike

◉ C. China

FEATURES

Male cones Pale yellow, borne in leaf axils

small female seed cone

Female cones Pale yellow, borne at shoot tips

Cephalotaxus fortunei (Cephalotaxaceae)

CHINESE PLUM YEW

This small evergreen was introduced into Europe from central China in 1849 by the Scottish plant collector Robert Fortune—hence the species name *fortunei*. It has a spreading habit, similar in outline to the common yew (p.37). In the wild in China, this tree grows naturally as a woodland species and as such is extremely shade tolerant. It is seldom cultivated in full sun.

The bark is red-brown and flaking, similar to that of the common yew. The Chinese plum yew's stems are erect with distinctive whorls of branches at intervals, each bearing ranks of linear, lanceolate leaves. Both male and female cones are small and pale yellow, and are produced on separate trees in spring. The female seed cone develops into an oval, blue-green, olive-like structure, up to ¾–1 ¼ in (2–3 cm) long. It matures to become purple-brown and plumlike over a period of two years. The fleshy covering is inedible and contains a single, small, brown nutlike seed.

pollen cone

linear, slightly sickle-shaped leaf with bright green upper surface

Leaves and cones
Male pollen cones of the Japanese plum yew are borne in leaf axils. The leaf undersides are gray-green.

FEATURES

yellow-green seed cone

Female cones Small; develop into fleshy structures

Cephalotaxus harringtonia (Cephalotaxaceae)

JAPANESE PLUM YEW

As the name suggests, this small conifer originates from Japan, where it has been cultivated for centuries, both as an ornamental evergreen and for its fruit. It is also native to Korea. It was introduced into the West in 1830 by German physician Philipp Franz von Siebold.

Also known as cow's tail pine because of its elegant drooping branchlets, Japanese plum yew grows as a spreading, bushy, shrublike tree, sometimes broader than tall. It has dense, spirally arranged foliage, which ascends from each branchlet in two planes to form a distinct V-shaped groove, rather like a bird's wings.

The bark is gray and peels in thin strips. Once pollinated, tiny female seed cones turn into light green oval seed structures, which mature to purple. They are fleshy and release a sweet-smelling cloudy resin when squeezed. In Japan, the ripe flesh and the seed inside are eaten, the latter having a pine-like flavor. If eaten before they are mature, they leave a pungent, resinous aftertaste, which may take hours to subside.

Chilean yew seed
Plumlike ovoid, edible seeds
develop from pollinated female
seed cones. Linear leaves are
spirally arranged around shoots.

*leaf twisted
on branch*

*leaf up to
1 in (2.5 cm) long*

*ovoid seed cone
¾ in (2 cm) long*

PROFILE

⬍ 22–80 ft (7–25 m)
↔ Up to 30 ft (10 m)
🕯 Evergreen
🌿 Needlelike
⊙ Chile, Argentina

Prumnopitys andina (Podocarpaceae)

CHILEAN YEW

This South American conifer, also known as
plum-fruited yew, is now considered vulnerable in
the wild. Native to the Andes mountains of southern
Chile and Argentina, it was introduced into cultivation
in Europe by English nurserymen Veitch & Sons in 1860.
It may reach a height of 100 ft (30 m), with a trunk up to
6 ft (2 m) wide in the wild. However, in cultivation it is a
small to medium-sized tree, somewhat resembling the
common yew (p.37) in habit.

The bark is graphite-gray. Leaves are bright green
on the upper surface and glaucous blue-green on the
underside. Each leaf has a distinct midrib. The Chilean
yew is dioecious, bearing male and female cones on
separate trees. Both male and female cones are small
and yellow-green. The yellowish brown seed cone
matures to blue-purple, and resembles a small damson.
Each seed is borne singly on a slender, scaly stalk. The
seed cone is eaten by indigenous Native Americans in
Chile, who also make marmalade from it.

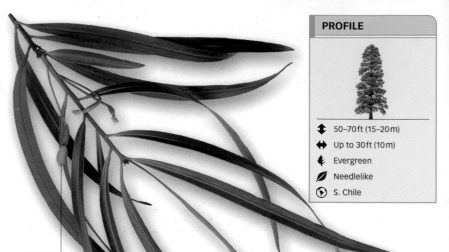

seed cone

narrowly linear, sickle-shaped leaf

pointed leaf, up to 6 in (15 cm) long

Lush foliage and cones
Drooping branches bear glossy, dark green leaves. Ovoid green seed cones, either solitary or in pairs, develop on the ends of long stalks.

FEATURES

Bark Attractive reddish brown in maturity

Male cones Greenish yellow; borne in clusters

Podocarpus salignus (Podocarpaceae)

WILLOW-LEAF PODOCARP

An extensively cultivated South American conifer, this tree is popular throughout the temperate world as an ornamental tree. However, the wild population in native Chile is now vulnerable due to habitat loss and excessive logging for its durable timber, which is used in construction work. Like a willow, this tree thrives on large amounts of rainfall.

The leaves have a pale underside, with visible bands of stomata (pores). A central ridge runs down each leaf. Greenish yellow male and female cones are borne on separate trees, with male pollen cones borne in clusters on terminal, slender spikes. The seed cone is up to ½ in (1 cm) long.

Willowy habit
Its leaf shape and frame, reminiscent of the willow, give willow-leaf podocarp its common name.

Cone clusters
Small male pollen cones form amid linear leaves. The leaves are arranged in two ranks along the shoots.

leaf up to 1 ¼ in (3 cm) long

tiny male cone cluster

slightly curved leaf has sharp tip

PROFILE

↕ 70–80 ft (20–25 m)

↔ Up to 30 ft (10 m)

🌿 Evergreen

▱ Linear

⊙ S. Chile, Argentina

FEATURES

Cones Male (bottom) and female (top) on same shoot

cone up to ½ in (1.5 cm) long

Seed cones Small, with fleshy scales

Saxegothaea conspicua (Podocarpaceae)

PRINCE ALBERT'S YEW

Both the generic and common names of this tree commemorate Prince Albert of Saxe-Coburg-Gotha, the consort of Britain's Queen Victoria. The genus *Saxegothaea* is monotypic, which means there is only one species in the genus. The tree is native to the temperate rainforests of southern Chile and Argentina, where it grows wild with other forest species such as Coigüe southern beech (p.160), Winter's bark (p.110), and *Podocarpus nubigenus*. All are valued for their timber. Prince Albert's yew was introduced into cultivation by Cornish plant collector William Lobb in 1847.

The linear leaves are similar to members of the *Taxus* (yew) genus, and dark green on the upper surface. Two distinct silvery bands of microscopic pores, or stomata, appear on the underside. Both male and female cones are small, and borne on the same tree. The seed is carried in a small cone with overlapping, pointed, green scales, borne singly at the end of twigs.

PROFILE

↕ 70–100 ft (20–30 m)

↔ Up to 70 ft (20 m)

❧ Evergreen

▱ Linear

◉ California

leaf twisted at base, arranged spirally on shoot

leaf up to 2 in (5 cm) long, lies flat on either side of shoot

two white bands on underside of leaf

Male cones
Numerous rows of yellow, male, pollen-bearing cones are borne in the leaf axils, from spring to early summer.

male pollen cone, up to ¼ in (7 mm) long

FEATURES

Bark Gray to gray-brown, with vertical ridges

Seed cones Green, up to 1½ in (4 cm) long; encase yellow-brown seeds

Torreya californica (Taxaceae)

CALIFORNIA NUTMEG

This tree is native to California, where it occurs sporadically in the Pacific Coast Ranges and the foothills of the Sierra Nevada. The California nutmeg has a conical outline, with a straight stem, and can grow large in the wild. Despite its name, it is not related to the true nutmeg (*Myrisitica fragrans*). It was introduced into western cultivation by plant collector William Lobb in 1851. At first glance, the California nutmeg looks similar to the *Cephalotaxus* genus (plum yews). However, it is easily distinguished by its stiff, sharply pointed leaves.

Mature shoots have a purplish hue. The male cones are about ¼ in (5–7 mm) long and grouped along the underside of each shoot, from where they disperse pollen. The female cones are borne on separate trees and mature into olive-shaped seed cones. The single, large seed found inside tastes like a peanut. The seeds were once a valued food of Native Americans.

PROFILE

- 50–70 ft (15–20 m)
- Up to 70 ft (20 m)
- Evergreen
- Needlelike
- W. Europe, E. Asia

Fleshy, fruit-like cones
The berrylike cone, or aril, is open at the top, exposing the seed. It ripens from green to red.

dark blackish green leaf

narrow leaf, up to 1¼ in (3 cm) long

young green aril with seed

leaves spread in two rows across shoot

FEATURES

Bark Rich brown; flaking to reveal pink patches

Male cones Small, pale yellow, pollen-bearing; borne beneath shoots

Taxus baccata (Taxaceae)

COMMON YEW

This evergreen tree has a natural range extending from western Europe into eastern Asia. In maturity, it develops a dense, dark canopy that gives it a somber feel. Mature trees are normally multi-stemmed and their spread may be as wide as their height. The common yew may live to a great age and some trees are believed to be several thousand years old. Yew wood is extremely durable and valued for veneer and furniture production.

Its leaves are slightly glossy on the upper surface, with a central groove, and a light green color on the underside. New growth is bright lime-green. Small male cones shed large amounts of pollen in spring. The soft, fleshy, berrylike cone (aril) encases an olive-green seed. All parts of the common yew are poisonous except for the aril.

♤ **Japanese yew (*Taxus cuspidata*)** This shrubby Asian yew can be distinguished from the common yew by its stiff, spine-tipped leaves, which are gold or brownish yellow on the underside.

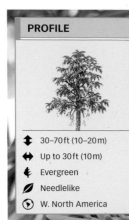

- ↕ 30–70 ft (10–20 m)
- ↔ Up to 30 ft (10 m)
- 🍂 Evergreen
- 🌿 Needlelike
- ⊙ W. North America

new leaf growth

Long-lasting foliage
Borne on shoots in two opposite, horizontally spreading rows, Pacific yew leaves remain on the tree for up to five years.

leaf up to 1 in (2.5 cm) long

FEATURES

Bark Thin, flaking; reddish brown in color

Taxus brevifolia (Taxaceae)

PACIFIC YEW

This small conifer is native to western North America, from British Columbia to California. It grows sporadically in canyons and gullies, often alongside streams. The anti-cancer drug paclitaxel (Taxol) was isolated from this tree's bark in the 1960s, and it has been used in the treatment of lung and breast cancer ever since. However, prolonged felling of the tree for the drug has led to concern among conservationists that it may become endangered in the wild. In recent years, paclitaxel has been produced semi-synthetically from cultivated Pacific yews, ensuring that the wild population is not further endangered.

Juvenile trees tend to be angular, becoming more conical with age. The branches are slender and slightly pendulous. The leaves are dark yellowish green on the upper surface and lighter on the underside. In winter, golden yellow scales cover the buds.

◊ **Chinese yew (*Taxus chinensis*)** Unlike Pacific yew, this tree has sparse foliage and slender, pale yellowish green, needlelike leaves that lie flat on either side of the shoot.

flattened spray of leaves

mid- to dark green foliage, shiny on top

open, woody, brown cone on branch tip

seed cone scales open wide

longer leaves at base of shoot

Foliage and cones
Dense foliage is carried in flattened sprays of overlapping scales. The seed cones have six scales and ripen to release up to four seeds.

PROFILE

- 70–160 ft (20–50 m)
- Up to 50 ft (15 m)
- Evergreen
- Scalelike
- W. United States, Mexico

FEATURES

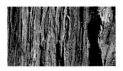

Bark Reddish brown; fissures in flaking, coarse plates

small golden cone

Cones Small male and female cones borne in small clusters at tips of shoots

Calocedrus decurrens (Cupressaceae)

INCENSE CEDAR

One of the most easily recognizable of all evergreen conifers, this tree is native to western North America, from Oregon to Baja California, Mexico. It has been widely cultivated elsewhere as an ornamental tree since its discovery in 1853. The name refers to the fragrance of the cut timber, which is durable and used for making chests, pencils, and roofing shingles.

The dense foliage extends down to within 6 ft (2 m) of the ground, even on elderly trees. The green shoots turn red-brown in maturity. Each tree carries small golden male and female cones. The seed cones, up to ¾ in (2 cm) long, are sometimes carried in such profusion that the foliage becomes pendulous under their weight.

Large, columnar tree
The incense cedar has a distinctive habit, with erect branches and a single, almost columnar form.

Foliage and cones
Male and female cones are borne on the same tree. Female cones develop from seed cones, about ¼ in (7 mm) wide.

dark green to blue-green leaf

young globular, green-purple seed cone

small, pointed leaf

PROFILE

- ⬍ Up to 130 ft (40 m)
- ⬌ 30–50 ft (10–15 m)
- Evergreen
- Scalelike
- S.W. Oregon to N.W. California

FEATURES

Bark Shiny gray-brown; vertically fissured in long plates

Male cones Produced at leaf tips; gray turning to red

wrinkled surface

Mature seed cones Woody, brown with 8–10 scales

Chamaecyparis lawsoniana (Cupressaceae)

LAWSON CYPRESS

This popular garden conifer is a large, conical tree with drooping branches, and makes an excellent hedge or screen. Originally a forest tree from the Pacific Northwest, where it remains an important timber-producing tree, its uses range from boatbuilding to cabinetmaking. The tree is named after Lawson's Nursery in Edinburgh, Scotland, which in 1854 was the first nursery in Europe to receive seeds from America.

The undersides of leaves are marked with white flecks where the scales meet. When crushed, the foliage has a distinctive smell of parsley. Male and female cones are borne in profusion on the same tree. Female cones are slate-blue, and borne on small branchlets away from the tips. Cone scales open as the seeds are shed, but the cone remains on the tree long after seed dispersal.

♠ **Taiwan cypress (*Chamaecyparis formosensis*)** Unlike the Lawson cypress, this tree has a slower growth habit and sharp pointed leaves, which make the sprays rough to the touch.

PROFILE

- 50–120 ft (15–36 m)
- Up to 30 ft (10 m)
- Evergreen
- Scalelike
- Cultivated hybrid, not found in the wild

leaves borne in a spray

small leaf with pointed tip

Spraying foliage
The Leyland cypress has small, scalelike leaves borne in flattened, slightly drooping sprays, irregularly arranged on shoots.

red-brown shoot

FEATURES

Bark Smooth, red-brown; develops shallow fissures in maturity

Female cones Angular, globular, up to ¾ in (2 cm) wide

x *Cupressocyparis leylandii* (Cupressaceae)

LEYLAND CYPRESS

Perhaps best known as a hardy, fast-growing conifer, the Leyland cypress is commonly planted as a garden hedge or screen. However, it is only suitable for planting in gardens if regularly trimmed and kept under control. This tree does not occur in the wild, being an intergeneric hybrid between two American species—the Monterey cypress (p.51) and the Nootka cypress (p.44). The original cross-pollination between the parent trees took place at Leighton Hall in Wales, United Kingdom, in 1888. The tree gets its name from Christopher Leyland, the property owner, who also planted the first seedlings in Britain. Several cultivars of this tree are commercially available.

The leaves have a dark green upper surface and a paler underside. They emit a pungent scent when crushed. Yellow male pollen cones and green female cones appear on the same tree, in clusters at the tips of shoots, in early spring. The mature seed-bearing cone is blue-green, ripening to woody brown. Each cone contains a single seed.

PROFILE

- 160–200 ft (50–60 m)
- Up to 50 ft (15 m)
- Evergreen
- Needlelike
- Japan, China

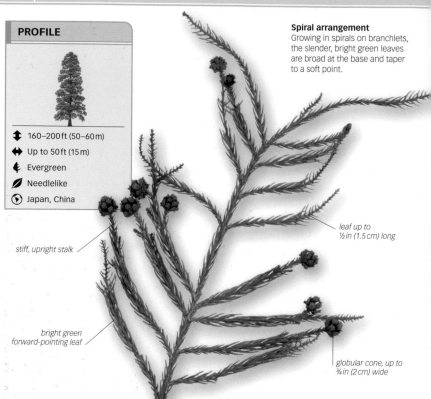

Spiral arrangement
Growing in spirals on branchlets, the slender, bright green leaves are broad at the base and taper to a soft point.

stiff, upright stalk

bright green forward-pointing leaf

leaf up to ½ in (1.5 cm) long

globular cone, up to ¾ in (2 cm) wide

FEATURES

Bark Soft, fibrous, red-brown; peels in long, vertical strips

female cones in cluster

Female cones
Green rosettes borne on stiff, upright stalks

Cryptomeria japonica (Cupressaceae)

JAPANESE CEDAR

This fast-growing conifer originates from Japan but has also been cultivated in China for centuries, where it has become naturalized. It first reached the West in 1842, when its seeds were sent to the Royal Botanic Gardens in Kew, London.

The curving branches of this tree sweep downward from the trunk, rising toward the tips. They may take root where they touch the ground. The foliage consists of bright green branchlets covered with stiff needles; at the base of each needle is a bulging keel (ridge) that runs down the branchlet. When crushed, the foliage emits an aroma resembling that of orange peel. Male and female cones are borne on the same tree. Yellowish, ovoid male pollen cones are clustered at the tips of branchlets, shedding pollen in early spring. The female seed cones develop into globular woody cones.

♤ *Taiwania cryptomeriodes* Whiplike branchlets and cylindrical cones, up to ½ in (1.2 cm) long, help to distinguish this tree from the Japanese cedar.

PROFILE

Fernlike foliage
This tree's thick, flattened sprays of foliage become pendulous towards the tips. When crushed, they emit a fruity, resinous scent.

flattened, dark green leaf

- ↕ 100–160 ft (30–50 m)
- ↔ Up to 40 ft (12 m)
- ⟊ Evergreen
- ⬙ Scalelike
- ◉ Japan

FEATURES

Bark Gray to red-brown, soft, fibrous; vertically fissured, peeling in narrow strips

Chamaecyparis obtusa (Cupressaceae)

HINOKI CYPRESS

This beautiful, broadly conical tree is native to Japan, where it is cultivated as an ornamental tree, as well as being important for timber. It is also held sacred by followers of Shinto Buddhism and often used in the construction of Shinto temples. The Hinoki cypress was introduced into the West in 1861.

The branches spread out almost horizontally from the trunk before dividing into purplish orange-brown shoots, which are covered in sprays of foliage. The leaves have a dark green upper surface and a pale yellow underside. Yellow male pollen cones are produced at the tips of each branchlet, releasing clouds of pollen in spring. Female cones are slate-blue, and carried on shorter branchlets on the same tree. They mature over two years into globular, plum-colored seed cones, up to ½ in (1 cm) long. Each woody cone scale carries a curved spine.

♧ **Sawara cypress (*Chamaecyparis pisifera*)** This tree has sharply pointed leaves, which have distinctive, white markings on the underside, and smaller cones than the Hinoki cypress.

PROFILE

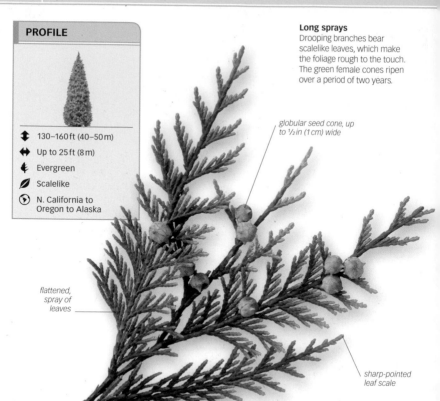

- 130–160 ft (40–50 m)
- Up to 25 ft (8 m)
- Evergreen
- Scalelike
- N. California to Oregon to Alaska

Long sprays
Drooping branches bear scalelike leaves, which make the foliage rough to the touch. The green female cones ripen over a period of two years.

globular seed cone, up to ½ in (1 cm) wide

flattened, spray of leaves

sharp-pointed leaf scale

FEATURES

Bark Pink-brown, vertically fissured; sometimes fluted toward the base

Chamaecyparis nootkatensis (Cupressaceae)

NOOTKA CYPRESS

This hardy North American species is a large forest tree with a broad, conical outline. It was discovered in 1793 by Scottish physician and naturalist Archibald Menzies, but not introduced into Europe until around 60 years later. It makes a handsome addition to parks, large gardens, and arboreta. Nootka refers to a historical (though inaccurate) European name for many Native American tribes that inhabited the northwest coast.

The leaves have a dark green upper surface and a yellow-green underside. Small male and female cones appear on the same tree from fall to spring. Male cones are yellow and carried on branch tips; the females are slate-blue.

◇ **White cypress**
This tree (*Chamaecyparis thyoides*) bears erect sprays of aromatic, glaucous-blue foliage and bloomy cones.

PROFILE

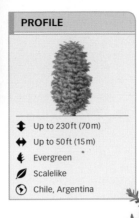

- Up to 230 ft (70 m)
- Up to 50 ft (15 m)
- Evergreen
- Scalelike
- Chile, Argentina

*seed cone
up to ½ in
(1 cm) wide*

*rounded,
brown
seed cone*

*leaf up to ⅛ in
(3 mm) long*

Oblong, thick leaves
The leaves of the Patagonian
cypress are bluntly pointed,
and carried in whorls of three
on pendulous shoots.

FEATURES

Bark Dark red-brown,
vertically fissured, peels
in shallow strips

Fitzroya cupressoides (Cupressaceae)

PATAGONIAN CYPRESS

A beautiful, large tree in its native habitat, the
Patagonian cypress is long-lived, with some specimens
living for more than 3,000 years. However, it is now
endangered in the wild due to forest clearance for
agricultural land. Discovered by Europeans traveling
with Charles Darwin on his 1834 expedition aboard HMS
Beagle, it takes its name from the ship's captain, Robert
Fitzroy. The plant collector William Lobb introduced it into
cultivation in 1849, and the tree is now planted in parks
and gardens, where it grows into a small tree with a
dense, cypress-like habit.

  This tree's trunk forks low to the ground, and the
bark peels to reveal pink new bark beneath. The upswept
branches end in pendulous sprays of blue-green foliage,
which is marked with two white bands of stomata (pores)
on each side. The same tree bears both male and female
cones in small clusters at the tips of the shoots in spring.
The seeds are held within lumpy cones, glossy green
ripening to brown, with scales opening wide.

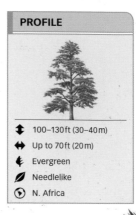

- ↕ 100–130 ft (30–40 m)
- ↔ Up to 70 ft (20 m)
- 🕭 Evergreen
- ⬤ Needlelike
- 🜨 N. Africa

Upright cones
Pollinated female seed cones develop on the top side of branches, ripening from green to brown.

barrel-shaped cone, up to 3 in (7.5 cm) long

leaves up to 1 in (2.5 cm) long

blue-green, needlelike leaf

leaves borne in whorls on side shoots

FEATURES

Bark Slate-gray; develops scaly plates when mature

male cone, up to 2 in (5 cm) long

Cones Golden male and green female cones, in separate clusters on same tree

Cedrus atlantica (Pinaceae)

ATLAS CEDAR

This conifer occurs naturally in the Atlas Mountains of North Africa, where it grows along the snowline around 7,200 ft (2,200 m) above sea level. Sometimes considered to be a geographical subspecies of the cedar of Lebanon (p.48), it is distinguished by its upswept branches. Its light blue form, *Cedrus atlantica* Glauca Group, is often seen in cultivation.

The leaves are borne in dense whorls on side-shoots and singly on leading shoots. The male cones disperse pollen in fall before falling to the ground. Pollinated female seed cones are green and borne in rows. After two to three years, the cones break apart on the tree, releasing winged seeds.

High rise
The Atlas cedar has upswept, ascending branches. The foliage varies from dark green to light gray-blue in the wild.

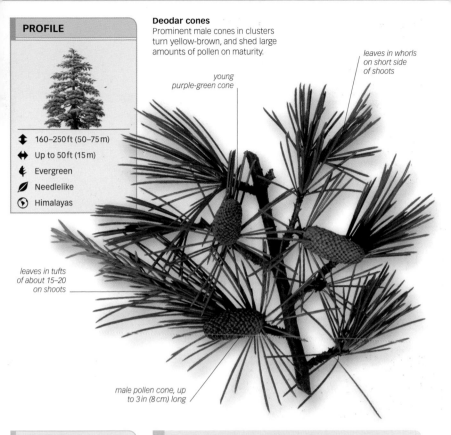

- 160–250 ft (50–75 m)
- Up to 50 ft (15 m)
- Evergreen
- Needlelike
- Himalayas

Deodar cones
Prominent male cones in clusters turn yellow-brown, and shed large amounts of pollen on maturity.

young purple-green cone

leaves in whorls on short side of shoots

leaves in tufts of about 15–20 on shoots

male pollen cone, up to 3 in (8 cm) long

FEATURES

Bark Smooth, dark gray; with age develops pink-gray fissures

Seed cones Resinous; growing up to 5 in (12 cm) long

Cedrus deodara (Pinaceae)

DEODAR CEDAR

Much admired across the temperate world, the deodar cedar is native to the western Himalayas including northern India. It is sometimes referred to as the Indian cedar. In the days of the British Raj, huge deodars—up to 250 ft (75 m) tall and close to 1,000 years old—could be found on hillsides surrounding the hill station of Simla. Most of these have been felled for their durable, light brown timber, which is highly prized for construction.

This tree is distinguished from other cedars by the drooping tips on its branches. Each shoot bears 1½ in- (4 cm-) long leaves that are blue-green when young, becoming dark green with age. The barrel-shaped seed cones are borne on the top of branches, green ripening to dull brown.

Distinctive habit
The graceful deodar tree is narrowly conical when young, broadening as it matures.

PROFILE

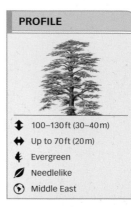

- ⬍ 100–130 ft (30–40 m)
- ⬌ Up to 70 ft (20 m)
- 🌿 Evergreen
- 🍃 Needlelike
- 🌐 Middle East

*whorls of rigid leaves
1 ¼ in (3 cm) long*

*erect male cone,
2 in (5 cm) long*

*flexible needles, up
to ¾ in (2 cm) long*

Cones and leaves
Male cones are upright and pale
gray-green. Leaves are arranged
in whorls on older shoots, but
appear singly on young shoots.

FEATURES

Female cones Barrel-shaped;
green turning to purple-brown

Cedrus libani (Pinaceae)

CEDAR OF LEBANON

Among one of the world's best known trees, the
cedar of Lebanon has been revered for thousands of
years, and in Biblical times was seen as a symbol of fertility.
King Solomon is believed to have built his temple using its
timber. On Mount Lebanon, it grows at altitudes of up
to 7,020 ft (2,140 m). It has been widely planted as an
ornamental species since the 17th century.

Narrowly conical when young, the tree broadens and
develops a flat-topped canopy made up of tiers of long,
horizontal branches as it matures. The bark is gray-brown,
with shallow, vertical fissures. On young shoots, leaves are
flexible and borne singly; older wood has up to 20 rigid
leaves arranged in whorls. Leaf color varies from
blue-gray to dark green. Ovoid female cones are pale
green, developing a rose-purple tint by early fall. Young
cones have dried, white resin on the scale margins.

♀ **Cyprian cedar (*Cedrus brevifolia*)** Although similar to the cedar
of Lebanon in many ways, the Cyprian cedar has greener and shorter
needlelike leaves, up to ½ in (1.2 cm) long.

PROFILE

- 100–145 ft (30–44 m)
- Up to 50 ft (15 m)
- Evergreen
- Needlelike
- California

*pointed, shiny
brown winter bud*

Widely spaced leaves
Stiff, pointed leaves of the
large-coned Douglas fir are
borne widely around the shoots.

*leaf up to 2 in
(5 cm) long*

FEATURES

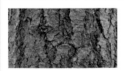

Bark Dull gray and smooth;
develops wide, vertical orange
fissures when mature

*paperlike
bract*

*cone up
to 7 in
(18 cm) long*

Cones Purple-green, cylindrical;
ripen to tobacco-brown

Pseudotsuga macrocarpa (Pinaceae)

LARGE-CONED DOUGLAS FIR

Sometimes known as bigcone fir, this large conifer
is native to southwest California, where some trees are
known to be over 600 years old. The overall form of this
tree is broadly conical, with frequent whorls of spreading,
slightly descending branches. It was introduced into
cultivation in Europe in 1910. It has the largest cones in
its genus, up to 7 in (18 cm) long and twice as big as its
better-known cousin, the Douglas fir (p.50).

The bark develops orange fissures with age. This
orange coloring also occurs on the shoots. These
are positioned toward the end of each branch, and are
covered with stiff needlelike leaves. The large cones
have numerous thick, woody scales. From beneath
each scale, a paperlike, trident-shaped bract protrudes.
The seeds inside each cone are large and heavy. Each
seed is attached to a short rounded wing, which is too
small to aid wind dispersal.

rich green leaf, up to 1 ¼ in (3 cm) long

two white bands on leaf's lower surface

PROFILE

- ↕ 300–330 ft (90–100 m)
- ↔ Up to 70 ft (20 m)
- 🌲 Evergreen
- ⬗ Needlelike
- ◉ British Columbia to California

cone up to 4 in (10 cm) long

three-pronged, papery bract overlying cone scale

Pendant cone
Long, hanging cones are a distinctive feature of this tree. The cones, green when young, turn orange-brown with age.

FEATURES

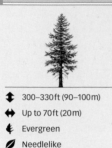

Bark Corky; vertically and deeply furrowed

immature seed cone

Seed cones Red, borne in clusters at end of shoot

Pseudotsuga menziesii (Pinaceae)

DOUGLAS FIR

This fast-growing conifer has a natural range from British Columbia in Canada to California. However, it has been planted throughout North America, Europe, Australia, and New Zealand for its commercially important timber. Discovered by Scottish physician Archibald Menzies, it was introduced into cultivation by another Scottish plant collector David Douglas. Both are commemorated in the tree's naming.

Mature trees have limited branching for the first 100 ft (30 m) with a long, straight trunk, which is chestnut-brown in color. Juvenile trees have smooth, shiny gray bark, marked with resin blisters and regular whorls of light, slightly ascending branches. The needlelike leaves are spirally arranged on shoots. When crushed, they emit a sweet, citrus fragrance. Winter buds are glossy, red-brown, and pointed.

◇ **Japanese Douglas fir (*Pseudotsuga japonica*)** This tree is distinguished from the Douglas fir by its hairless shoots, smaller cones, and shorter, needlelike leaves.

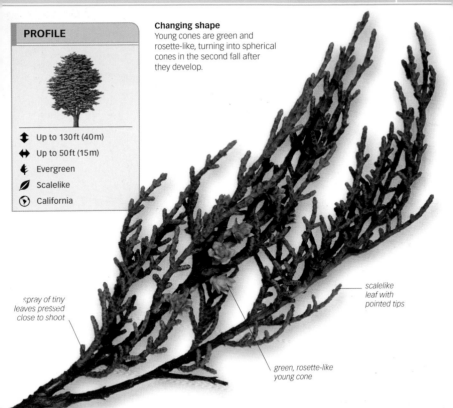

Changing shape
Young cones are green and
rosette-like, turning into spherical
cones in the second fall after
they develop.

↕ Up to 130ft (40m)

↔ Up to 50ft (15m)

🌡 Evergreen

🌿 Scalelike

⦿ California

spray of tiny
leaves pressed
close to shoot

scalelike
leaf with
pointed tips

green, rosette-like
young cone

FEATURES

Bark Peels in vertical strips
when mature to reveal fresh
pink-brown bark

cone up
to 1½in
(3.5cm) wide

Cones Spherical, shiny purple-
brown when ripe

Cupressus macrocarpa (Cupressaceae)

MONTEREY CYPRESS

This large conifer originates from the coastal
cliffs of the Monterey Peninsula, California. It can
withstand exposure to salt-laden winds and has been
widely planted to provide shelter in coastal
regions in temperate zones. It has a flat-topped,
spreading crown of large, heavy, horizontal branches,
reminiscent of the cedar of Lebanon (p.48).

Young trees are fast-growing, with a conical
or broadly columnar habit. They have lots of
upswept branches with spiky shoots and leaves
that turn dark green with age. The pale brown bark
has fine shallow ridges. When crushed, the foliage
emits a lemony fragrance. The male cones release
clouds of dustlike pollen in early summer.

♤ **White cedar (*Cupressus lusitanica*)** Smaller and less hardy
than the Monterey cypress, this graceful tree has pendulous
branches and a brown peeling bark.

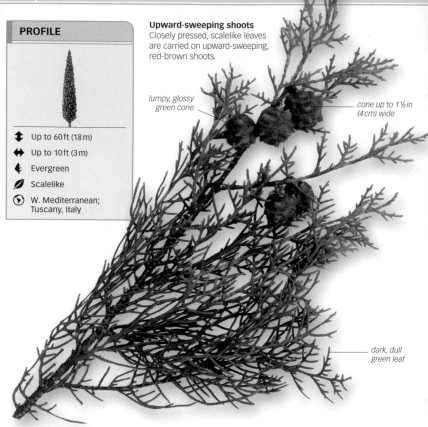

PROFILE

Up to 60 ft (18 m)

Up to 10 ft (3 m)

Evergreen

Scalelike

W. Mediterranean;
Tuscany, Italy

Upward-sweeping shoots
Closely pressed, scalelike leaves
are carried on upward-sweeping,
red-brown shoots.

*lumpy, glossy
green cone*

*cone up to 1½ in
(4 cm) wide*

*dark, dull
green leaf*

FEATURES

Bark Brown-gray; with shallow,
scaling ridges visible toward
the base of the trunk

Cupressus sempervirens (Cupressaceae)

ITALIAN CYPRESS

Also called the Mediterranean cypress, this is one
of the most easily recognizable evergreen conifers. It
is widely found in the hills, farmland, and roadsides of
Tuscany, Italy, and the western Mediterranean region. The
tree has a distinctive tight, narrowly columnar form, which
it retains all through its life. It has been grown as an
ornamental tree, to soften and accentuate architecture,
since the rise of the Greek civilization.

Ascending branches covered with dense foliage
obscure the stem from around head height. The
foliage, unusually for cypresses, has little noticeable
scent when crushed. The cones are shiny green, turning
dark red-brown, and finally gray, staying on the tree for
many years. They are larger than the cones of most other
cypresses and found spread across the crown.

♀ **Arizona cypress (*Cupressus arizonica*)** This tree has a
densely conical form, unlike the distinctive narrow columnar shape
of its Mediterranean cousin.

PROFILE

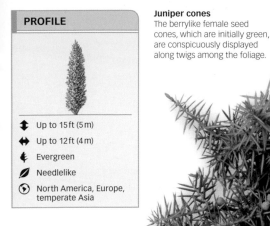

- ⬍ Up to 15 ft (5 m)
- ⬌ Up to 12 ft (4 m)
- 🕯 Evergreen
- Ø Needlelike
- ⊙ North America, Europe, temperate Asia

Juniper cones
The berrylike female seed cones, which are initially green, are conspicuously displayed along twigs among the foliage.

needle up to ½ in (1 cm) long

gray-green, awl-shaped leaf

leaves arranged in dense whorls of three

silver-backed needle with sharp, prickly point

female seed cone, up to ¼ in (6 mm) long

FEATURES

Bark Dark red-brown, peeling in vertical papery strips

frosty, powdery bloom

Female cones Berrylike; ripen to blue-black over two to three years

Juniperus communis (Cupressaceae)

COMMON JUNIPER

The common juniper is believed to be the most widespread tree in the world. Its natural range extends from Alaska, Greenland, Iceland, and Siberia southward through most of Europe, temperate Asia, and North America. It is a small, hardy tree that can tolerate intense cold, high elevations in mountainous terrain, and exposed coastal locations.

New shoots are blue-green with some "blooming," becoming dark red-brown with age. Male and female cones are yellow. They are borne on separate trees in clusters within the leaf axils. Berrylike female seed cones, called juniper berries, are used to give gin its characteristic flavor.

♤ Alligator juniper
Unlike the common juniper, this tree (*Juniperus deppeana*) has silver-blue leaves and bark that resembles alligator skin.

Male cones and leaves
Pollen-bearing cones are borne from fall until early spring, when the pollen is shed. Juvenile leaves are awl-shaped.

mature, scalelike leaf

yellow male cone carried at foliage tips

PROFILE

↕	Up to 70 ft (20 m)
↔	Up to 15 ft (5 m)
🌿	Evergreen
▰	Scalelike
⊙	N. and E. China; Japan, Mongolia

FEATURES

Bark Gray to reddish brown; peeling off in strips

Juniperus chinensis (Cupressaceae)

CHINESE JUNIPER

Since its introduction in the West in 1767, many different garden cultivars have arisen from the Chinese juniper. It is also one of the most common conifers used for bonsai (the cultivation of miniature versions). This is an extremely variable species in the wild, growing in north and east China, Mongolia, and coastal regions of Japan.

In its typical form, the tree is compact, erect, columnar, or conical. The bark is sometimes deeply fluted at the base of the trunk. Juvenile leaves are sharp, awl-shaped, up to ½ in (1 cm) long, and rigid. On mature trees, the leaves turn dark green with glaucous blue banding on the underside. When crushed, the foliage emits a "catty" smell. Small male and female cones are scalelike, yellow, and borne on separate trees. The seed cones grow up to ¼ in (7 mm) long, green ripening to bright glaucous white in the second year.

◊ **Drooping juniper (*Juniperus recurva*)** This Himalayan juniper can be distinguished from the Chinese juniper by its shaggy gray bark and its attractive, drooping branchlets.

sharply pointed
leaf, up to ⅜ in
(9 mm) long

pale green,
awl-shaped leaf

Dense foliage
The leaves are arranged on shoots
in six ranks of alternating whorls
of three. They range in color from
pale green to a steely blue-green.

PROFILE

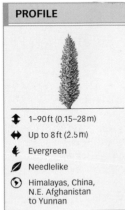

- ↕ 1–90 ft (0.15–28 m)
- ↔ Up to 8 ft (2.5 m)
- Evergreen
- Needlelike
- Himalayas, China, N.E. Afghanistan to Yunnan

Juniperus squamata (Cupressaceae)

SCALY-LEAVED NEPAL JUNIPER

Also known as flaky juniper or Himalayan juniper, this slow-growing tree ranges from a prostrate shrub to a small, bushy tree. Perhaps surprisingly, this untidy-looking species is the parent of many popular garden cultivars, several of which are grown as dwarf conifers. All forms have characteristic nodding tips to the shoots, which is a distinct feature of this species.

The red-brown, papery bark flakes in large scales. Male and female cones are usually borne on separate trees. Yellow-green male cones, up to ⅛ in (4 mm) long, shed pollen in early spring. Female cones develop into globular to ovoid, berrylike structures, up to ⅜ in (9 mm) long. Initially reddish brown, they become glossy purple-black when mature. Each cone contains a single seed.

♤ **Prickly juniper (*Juniperus oxycedrus*)** This related tree is distinguished by its open, drooping habit and rigid leaves with two white bands of stomata (pores) on the undersides.

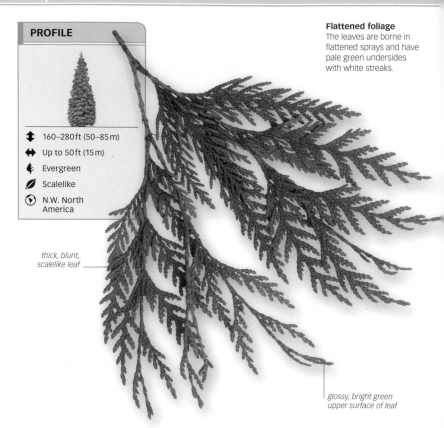

Flattened foliage
The leaves are borne in
flattened sprays and have
pale green undersides
with white streaks.

*thick, blunt,
scalelike leaf*

*glossy, bright green
upper surface of leaf*

FEATURES

Bark Red-brown, fibrous;
shreds in maturity

Female cones Small, ripening
from green to brown; open
on branchlets

Thuja plicata (Cupressaceae)

WESTERN RED CEDAR

This large, fast-growing conifer is found along the
northwestern Pacific coastline of North America, where
it is a major component of the lowland coniferous forests.
In these forests, specimens of western red cedar more
than 1,000 years old have been recorded. These have vast,
buttressed trunks and multiple stems, which are very often
layers from the original tree. The tree produces strong,
durable, red timber, which has been used over centuries to
build everything from Native American canoes and roofing
shingles to garden greenhouses.

Curved, sweeping branches carry large, drooping sprays
of foliage. When crushed, the foliage emits a pleasant,
fruity aroma, like that of a pineapple. Small, pale yellow
male pollen cones are borne at the tips of shoots. Small
female cones develop into leathery, erect ovoids with
rounded scale tips, ½ in (1 cm) long.

♤ **Pahautea (*Libocedrus bidwillii*)** In addition to its upright,
columnar habit, pahautea is distinguished from the western red
cedar by its smaller leaves and cones.

Upright sprays
The twisting sprays of mid-green foliage are gray-yellow on the underside. They emit the scent of fresh green apples when crushed.

leaf ¹⁄₁₆ – ¼ in (2–5 mm) long

flat shoot

PROFILE

↕ Up to 70 ft (20 m)
↔ Up to 43 ft (13 m)
❅ Evergreen
⬮ Scalelike
◉ E. North America

FEATURES

cone opens to base of scales

Seed cones Yellow; becoming brown and woody when mature

Thuja occidentalis (Cupressaceae)

WHITE CEDAR

Also known as American arbor-vitae, this hardy, slow-growing conifer has a columnar habit. Found at high altitudes in forests on rocky outcrops in eastern North America, it is an important source of timber in the United States. It is thought to be the first North American tree to have been introduced into Europe, sometime between 1534 and 1536. Since then it has been widely cultivated in parks and gardens. It has given rise to many cultivars, mostly slow-growing dwarf varieties, which are often grown in ornamental rockeries.

The spreading branches are upcurved at the tips. The reddish brown bark peels in vertical strips. New cones are borne in abundance on the top side of the foliage, giving the whole crown a golden appearance.

🜨 **Chilean cedar**
This tree (*Austrocedrus chilensis*) can be recognized by its sea-green, fernlike, flattened foliage.

*cone up to ½ in
(1 cm) wide*

Frond-like foliage
Erect branches bear upright sprays
of foliage made up of thick, blunt
scales, which are vibrant green on
both upper and lower surfaces.

*green, blunt,
scalelike leaf*

PROFILE

↕	70–100 ft (20–30 m)
↔	Up to 15 ft (5 m)
🍂	Evergreen
🌿	Scalelike
◉	China, Korea

Platycladus orientalis (Cupressaceae)

CHINESE ARBOR-VITAE

Until recently the botanical name for this tree
was *Thuja orientalis* and some references still name
it as such. Originating from China and Korea, this
attractive, small tree produces a dense conical or
columnar habit, especially when young. Old trees
often display bare, upcurved branches high in the
crown. The word *arbor-vitae* is Latin for "tree of life."
In Buddhist China, the tree is associated with long
life and vitality, perhaps due to its unchanging
evergreen habit, and the fact that several of the
trees planted around Chinese Buddhist temples
are more than 1,000 years old.

Chinese arbor-vitae has matte red-brown bark,
which shreds in vertical strips. The cones are very
small, dull yellow, and borne on the tips of branchlets.
Female seed cones are ovoid, borne upright, and
a glaucous-green color ripening to brown. Several
garden cultivars have been developed.

*leaf bright
yellow-green on
upper surface*

Glossy foliage
Hiba's foliage is so glossy that
from a distance it looks artificial,
almost like plastic.

*leaf up to ¼ in
(6 mm) long*

FEATURES

Female cones Ovoid, up to
¾ in (2 cm) long; green ripening
to woody brown

Thujopsis dolabrata (Cupressaceae)

HIBA

This evergreen tree is the only species in its genus.
It can be distinguished from the similar-looking *Thuja*
species by its broader leaves, each with striking and
distinctive white banding on the underside. It was
introduced into western cultivation in 1853.

A slow-growing tree, hiba takes on a dense, shrublike
form in the first 10 years or so. In maturity, it may have
a tall, upright, and columnar habit or a low-branching,
multi-stemmed, and broadly conical one. The red-brown
to gray bark peels off in strips, giving it a shredded
appearance. The foliage is made up of hard, scalelike
leaves that form large flattened sprays. Their upper
surface is bright, glossy yellow-green, while the
underside is white, with green margins. Both male and
female cones are borne in spring, in separate clusters
on the same tree. The blackish green male pollen cones
appear at the tips of the branchlets. Blue-gray female
seed cones develop into woody, ovoid cones that ripen
from green to brown.

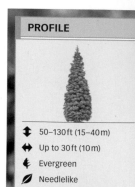

- 50–130ft (15–40m)
- Up to 30ft (10m)
- Evergreen
- Needlelike
- Tasmania

sharp-edged leaf curving inward

rigid, triangular, overlapping leaf

ovoid cone, approximately ¾in (2cm) wide

Leaves and cones
The foliage is made up of bright green, clawlike, leathery leaves. The cones ripen from bright green to orange-brown.

FEATURES

Bark Soft, rich red-brown, turning gray-brown with age

Athrotaxus selaginoides (Cupressaceae)

KING WILLIAM PINE

This medium-sized, evergreen tree is sometimes called the Tasmanian cedar, or King Billy pine. However, it is neither a cedar nor a pine, being related to cypress instead. It is found in Tasmania, where it grows at an altitude of up to 3,940ft (1,200m) above sea level alongside *Eucalyptus*, in mountain forests that are frequently covered in snow. It is considered to be threatened in the wild, with almost half of its natural range destroyed by bush fires started by loggers clearing debris after felling *Eucalyptus* trees.

From a distance, this sprawling tree could be mistaken for the Japanese cedar (p.42). Up close, the bark, which is soft and fibrous, is sometimes compared to that of the giant redwood (p.63). The foliage is made of short, spike-shaped needles, which curve away from bright apple-green shoots. Cones are bright green at first and ripen to orange-brown. They are borne in pairs on 1 in- (2.5 cm-) long stalks. The wood is used to make musical instruments and furniture.

Leaves and cones
When young, the shoots and leaves are pale yellow and soft. Yellow-green cones are borne at the end of shoots.

young leaf, up to ¼in (0.5cm) long

bract overlying cone scale

tip of leaf curves inward at the tip

↕ 30–70ft (10–20m)

↔ Up to 30ft (10m)

🌿 Evergreen

▨ Scalelike

◉ W. Tasmania

deep green, rigid, mature foliage

ovoid female cone, up to ¾in (2cm) wide

FEATURES

Bark Coppery, becoming deeply fissured with age

Athrotaxus laxifolia (Cupressaceae)

SUMMIT CEDAR

A small to medium-sized, hardy tree, the summit cedar differs from the King William pine (p.60) in its more lax habit, leading to its botanical name. Native to western Tasmania, where it grows in high mountain ranges, this tree is able to withstand temperatures of up to 10.4°F (–12°C) without damage.

The tree has pale red-brown or coppery bark. its triangular, pointed, scalelike leaves spread away from the shoots. Yellow-brown male and female cones are borne on the same tree in early spring. Tiny male pollen cones are clustered together at the tips of branchlets, while female seed cones, sometimes solitary, are borne at the ends of shoots. Bright yellow-green, they ripen in fall.

Lax habit
The summit cedar's broadly conical crown develops into a rounded, open top as the tree matures.

Cones on branches
Rounded seed cones form on the tips of the branches, against a background of dark green, linear leaves.

leaf up to ¾ in (2 cm) long

leaves in two flat rows on shoots

cone ¾–1¼ in (2–3 cm) long

↕ 130–360 ft (40–110 m)

↔ Up to 75 ft (23 m)

🗲 Evergreen

🌿 Needlelike

⊙ S.W. Oregon, N.W. California

FEATURES

Bark Red-brown, thick, soft, and fibrous

Seed cones Pendant, rounded; green ripening to brown

Sequoia sempervirens (Cupressaceae)

COASTAL REDWOOD

This magnificent species bears the distinction of producing the tallest tree in the world—currently 380 ft (115 m) tall. It is named after the Cherokee *sequoia*. Its natural range is limited to California and Oregon, where it grows on the coastal side of the Sierra Nevada mountains. From here, it was introduced into Europe in 1840.

Young trees are conical, with widely spaced, slender branches that curve upward at the tips. The leaves are speckled with two bands of white stomata (pores) on the underside. Yellow-brown male and green female cones appear in separate clusters on the same tree. The female seed cones turn brown and woody when ripe.

Standing tall
Mature coastal redwood trees acquire a columnar habit, with flat tops and downswept branches.

PROFILE

- 160–310ft (50–95m)
- Up to 50ft (15m)
- Evergreen
- Scalelike
- California

deep green leaf ¼in–½in (6–12mm) long

sharp-pointed, awl-shaped leaves

leaves arranged spirally around branchlets

male pollen cone

Giant redwood sprig
Male pollen cones develop at the tips of minor shoots and shed clouds of yellow pollen in spring.

FEATURES

Bark Red-brown, spongy and fibrous, up to 20in (50cm) thick

one-year-old cone still green

Seed cones Long-stalked, ovoid, up to 2in (5cm) wide

Sequoiadendron giganteum (Cupressaceae)

GIANT REDWOOD

The largest living tree is a giant redwood called "General Sherman," which grows on the western slopes of the Sierra Nevada in California. Giant redwoods are long-lived, with some specimens believed to be 3,500 years old. Introduced into Europe in 1853, it was named "Wellingtonia" to commemorate the British Duke of Wellington, who died in 1852.

The tree has large, downswept branches and a clear trunk of many feet from the ground. The leaves emit a smell of aniseed when crushed. Erect, bright green female cones develop into bunches of drooping, green seed cones. These ripen in two years, becoming brown and woody in the process.

Tree giant
Weighing up to 2,200 tons, giant redwood has a conical crown that turns broadly columnar with age.

- ↕ 160–280ft (50–80m)
- ↔ Up to 70ft (20m)
- ✦ Evergreen
- ⬤ Needlelike
- ◉ W. North America

Foliage and cones
The dark green leaves are bright lime-green when young. The cones mature from purplish red to deep brown, with papery scales.

leaf up to ¾in (2cm) long

purplish red young cone

FEATURES

Female cones Small, red; grow on shoot tips

Mature seed cones Woody, egg-shaped; 1in (2.5cm) long

Tsuga heterophylla (Pinaceae)

WESTERN HEMLOCK

Native to the west coast of North America, this tall, elegant conifer is characterized by its spreading, arching branches and soft pendulous foliage. This hardy, fast-growing species is shade-tolerant and capable of outgrowing its competitors in dense forests. It produces strong timber, and since its discovery in 1828 by Scottish botanist David Douglas, and introduction into cultivation in 1851, it has been widely grown in plantations as a timber-producing tree in North America and Europe.

This tree has a narrow, conical habit, with a lax leading shoot and thin, ascending branches that arch downward toward the tip. The bark is reddish purple in young trees, turning dark purple-brown in maturity. The leaves have two blue-white bands of microscopic pores, or stomata, on the underside. The male pollen cones are red and release large amounts of pollen in spring.

♤ **Mountain hemlock (*Tsuga mertensiana*)** This related species is easily distinguished from the western hemlock by its oblong-cylindrical large cones and radially arranged leaves.

- Up to 100 ft (30 m)
- Up to 50 ft (15 m)
- Evergreen
- Needlelike
- E. North America

vibrant green leaf, up to ¾ in (2 cm) long

needle-shaped leaf, with blunt tip

Eastern hemlock cones
Female cones, up to ½ in (1.5 cm) long, are borne on the tips of shoots. Green at first, the cones turn coffee-brown when mature.

FEATURES

Bark Brown-gray; turns scaly and fissured with age

young seed cone

Cones Small, oval; borne on short stalks

Tsuga canadensis (Pinaceae)

EASTERN HEMLOCK

This large tree is sometimes called the Canadian hemlock. Unlike its cousin, the western hemlock (p.64), it may grow either as a multiple-stemmed tree, or with heavy branching from near ground level. It is native to eastern North America, where it is currently under threat from an insect known as the hemlock woolly adelgid, a sap-sucking bug accidentally introduced from eastern Asia in 1924.

Branchlets of this tree vary in color, from yellow brown to almost cream, and bear small, green-brown, ovoid winter buds at the tips. The leaves are fresh green at first, becoming dark green with age. They have narrow, silvery bands on the lower surface. Yellow-green male pollen cones are borne in clusters along shoots, shedding pollen in spring. Female cones are green and borne in separate clusters from males.

◊ **Japanese hemlock (*Tsuga sieboldii*)** Except for its single-stemmed, spire-like habit and larger cones, up to 1¼ in (3 cm) long, this tree is similar in most aspects to the eastern hemlock.

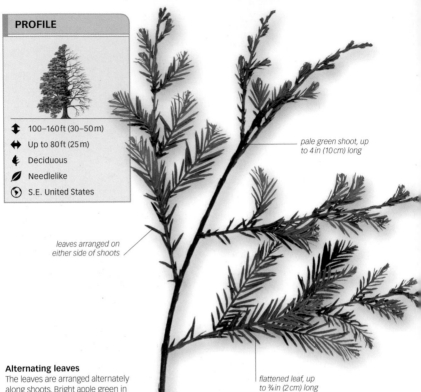

pale green shoot, up to 4 in (10 cm) long

leaves arranged on either side of shoots

flattened leaf, up to ¾ in (2 cm) long

Alternating leaves
The leaves are arranged alternately along shoots. Bright apple green in late spring, they turn plum-colored or orange-brown in fall.

FEATURES

Bark Dull reddish brown, fibrous and frequently fluted

Aerial roots Rise above water level to absorb oxygen

Taxodium distichum (Cupressaceae)

SWAMP CYPRESS

Also known as bald cypress, this ornamental conifer is native to the southeastern United States, where it grows in damp or wet soils. It is widespread in the Florida Everglades. This tree tolerates having its roots submerged in water for several months at a time by developing aerial roots known as "knees" or "pneumatophores."

The crown is conical, becoming slightly domed in maturity with heavy, upswept branches. The base of the trunk may be strongly buttressed. Both male and female cones occur on the same tree. In winter, male pollen cones appear in catkin-like clusters at the end of shoots. They grow up to 4–12 in (10–30 cm) long before shedding pollen in spring. Small, green female seed cones ripen into short-stalked, globular, pale brown cones.

◌ **Pond cypress (*Taxodium distichum* var. *imbricarium*)**
Smaller and slower-growing, this cypress has spreading branches that bear erect shoots and foliage.

PROFILE

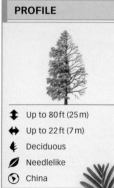

- ↕ Up to 80 ft (25 m)
- ↔ Up to 22 ft (7 m)
- 🍂 Deciduous
- 🌿 Needlelike
- 🕐 China

flattened leaf,
down-curved at tips

Feathery leaves
Leaves are soft at first, almost feathery, and are borne in two opposite ranks on the shoot. A pale band runs down each side of the midrib on the leaf underside.

leaf up to ¾ in (2 cm)
long, bright green on
upper surface

FEATURES

Bark Bright cinnamon-brown; peels in stringy, vertical flakes

cone ¾ in
(2 cm) wide

stalk ¾ in
(2 cm) long

Seed cones Cylindrical; green maturing to dark brown

Metasequoia glyptostroboides (Cupressaceae)

DAWN REDWOOD

Also referred to as the fossil tree, this conifer was discovered in 1941 by Chinese forester T. Kan in eastern Sichuan province. Until then, it had only been seen as a fossil and was believed to have been extinct for many millennia. It was introduced into western cultivation in 1948. Since then the tree has become a popular species for planting in parks and gardens.

The fibrous bark sometimes has a spiraled appearance. The trunk is fluted with buttressing at the base. The bright green leaves turn pink, sometimes gold, in fall before being shed. Male and female cones are borne on the same tree, with male pollen cones held in prominent clusters.

Upswept branches
Conical at first, the crown of the dawn redwood may become broad in maturity, with upswept branches.

leaf up to 2½ in
(6 cm) long

Spirally set leaves
Although spirally set, the needlelike
leaves twist to lie in two planes on
either side of the shoots.

*mature male cones
in a cluster*

PROFILE

⬍ Up to 70 ft (20 m)

⬌ Up to 30 ft (10 m)

🌢 Evergreen

▰ Needlelike

◉ S. and W. China

FEATURES

Bark Chestnut-brown, with
long, parallel, vertical fissures

Seed cones Rounded; bright
green maturing to brown

Cunninghamia lanceolata (Cupressaceae)

CHINESE FIR

At first glance, the Chinese fir may seem similar to
the monkey puzzle tree (p.28). However, it is a member
of the cypress family and not a true fir as the name
suggests. It is a handsome conifer with a domed crown
of short, drooping branches made up of distinctive,
glossy, dark green foliage. It has become popular as
an ornamental tree in parks and gardens.

Widespread branches give this tree an open and
somewhat sparse appearance, but this is often masked
by its drooping foliage. Its large, prickly leaves are glossy
green on the upper surface, with two distinctive bands
of white stomata (pores) on the underside. The same
tree bears both male and female cones. Male cones are
yellow-brown and borne in clusters at the shoot tips, while
female cones are yellow-green and held singly on terminal
branches. The female cone develops into a squat, rounded
seed cone, up to 1½ in (4 cm) wide.

🌢 **Luanta fir (*Cunninghamia konishii*)** This is a smaller tree than
the Chinese fir and has smaller, needlelike leaves and cones.

Whorled leaves
The leaves are borne in whorls, which are set 1½ in (4 cm) apart, on buff brown shoots. Yellow-green, globular male pollen cones are borne in clusters of 12.

shiny, rich green, needlelike leaves, up to 5 in (12 cm) long

egg-shaped cone, up to 3 in (7.5 cm) wide

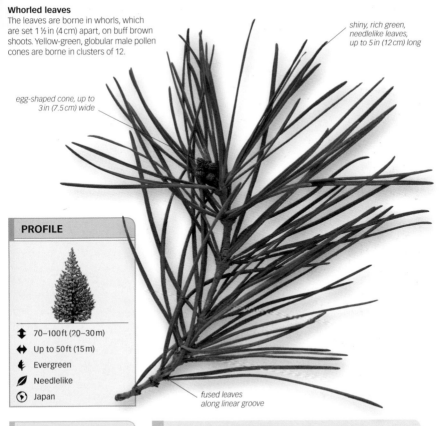

PROFILE

↕ 70–100 ft (20–30 m)

↔ Up to 50 ft (15 m)

🌿 Evergreen

🍃 Needlelike

⊙ Japan

fused leaves along linear groove

FEATURES

Bark Dark brown to gray; peels in long coarse strips

Sciadopitys verticillata (Sciadopityaceae)

JAPANESE UMBRELLA PINE

This distinctive tree is not a pine but the single member of a conifer family known in fossil records for 230 million years. Native to the mountains of central Honshu in Japan, it grows at elevations of up to 5,000 ft (1,520 m) on rocky slopes and ridges.

The leaves are borne in umbrella-like whorls, hence the tree's common name. Immature green female cones develop at the ends of each shoot. Mature seed cones are green, turning to woody brown. They have fleshy grooved scales when young. Needles may sometimes grow out from the center of the cone.

Attractive conifer
Japanese umbrella pine is widely planted because of its unusual leaf formation and regular shape.

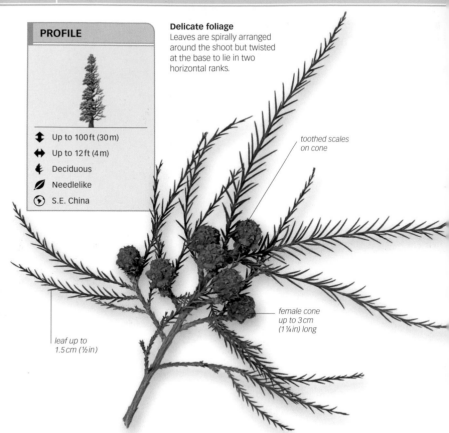

PROFILE

- Up to 100ft (30m)
- Up to 12ft (4m)
- Deciduous
- Needlelike
- S.E. China

Delicate foliage
Leaves are spirally arranged around the shoot but twisted at the base to lie in two horizontal ranks.

toothed scales on cone

female cone up to 3cm (1¼in) long

leaf up to 1.5cm (½in)

FEATURES

Bark Gray-brown with shallow fissures

Glyptostrobus pensilis (Cupressaceae)

WATER PINE

This conifer is also known as the Chinese swamp cypress. It is indigenous to southeastern China, where it grows naturally in swamps and along riverbanks. However, the wild population is now endangered as a result of overcutting for its durable, scented wood. Elsewhere, it has been extensively planted along the banks of rice paddies, where its root system helps to stabilize banks and reduce soil erosion. Despite its common names, it is neither a pine nor a cypress, but a monotypic (single species) genus related to *Taxodium*. Like that genus it produces "knees," or aerial roots, when planted directly in water.

The water pine has sea-green, soft, deciduous leaves. The shoots are gray-green, sometimes angular, and rough to the touch due to "pegs" (raised scars) left by discarded leaves. In fall, the leaves turn purple-brown before being shed. The cones are small and visually insignificant. The female seed cone is rough and pear-shaped. Green at first, it ripens to yellow-brown when each scale opens wide to release winged seeds.

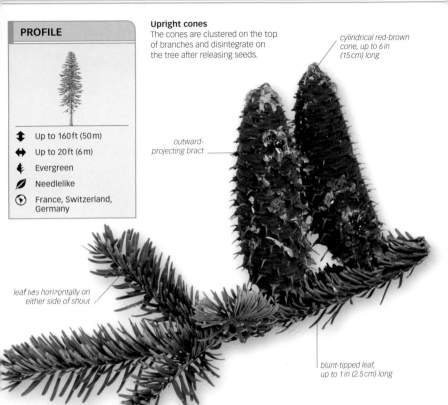

Up to 160 ft (50 m)

Up to 20 ft (6 m)

Evergreen

Needlelike

France, Switzerland, Germany

Upright cones
The cones are clustered on the top of branches and disintegrate on the tree after releasing seeds.

cylindrical red-brown cone, up to 6 in (15 cm) long

outward-projecting bract

leaf lies horizontally on either side of shoot

blunt-tipped leaf, up to 1 in (2.5 cm) long

FEATURES

Leaves Dark green, two silver bands on the underside

Male cones Greenish yellow; grouped on the undersides of shoots

Abies alba (Pinaceae)

EUROPEAN SILVER FIR

This large, evergreen conifer is a common species in central Europe, particularly in the alpine mountains of France, Switzerland, and Germany. It is relatively long-lived for a conifer, with some specimens known to be over 300 years old. The silver fir is used as a Christmas tree in many parts of Europe. It is also widely planted for timber, although it needs to be guarded against aphid attacks, which can prove fatal.

Young trees have a narrowly conical and symmetrical form, with slightly ascending branches. When mature, they become forked and more heavily branched, with dead branches covering the trunk for the lower third of its height. The bark is smooth and dull gray on young trees, becoming paler and developing shallow pink-brown fissures as they age. The leaves have white bands on the underside, which make them appear silvery.

 Greek fir (*Abies cephalonica*) Unlike the European silver fir, this large, handsome tree has stiff, sharp-pointed leaves that are arranged radially around the shoot.

- Up to 130 ft (40 m)
- Up to 15 ft (5 m)
- Evergreen
- Needlelike
- Caucasus, N.E. Turkey

Foliage and cones
The leaves are flat needles with blunt tips, borne more densely on the upper sides of shoots. The seed is contained within a brown, upright, cylindrical cone.

round-tipped leaf, up to 1 in (2.5 cm) long

cone 6–8 in (15–20 cm) long

male pollen cone opens yellow

Male cones Red-tinged; borne in clusters below the shoot

Abies nordmanniana (Pinaceae)

CAUCASIAN FIR

This relatively slow-growing conifer has dense branches and a uniformly conical outline, which makes it a magnificent specimen in parks, gardens, and arboreta. Since its introduction into western cultivation in 1848, it has been widely planted outside its native regions. It is also a popular Christmas tree and, under the name of Nordmann fir, is marketed as a "non-drop" tree, since it retains its needles for some time after being cut. When crushed, the foliage emits a fruity, resinous aroma.

The bark is gray and smooth, cracking into small, square plates in maturity. Glossy dark green on the upper surface, the leaves have two white bands of stomata (pores) on the underside. Male and female cones are borne on the same tree. The female seed cone, carried above the shoot, is green and turns red-brown as it matures. It has papery bracts, which project beyond each scale.

🔊 **Algerian fir (*Abies numidica*)** This tree is distinguished from the Caucasian fir by its more downward-curving branches, as well as its densely set, thick, rounded, needlelike leaves.

cone scale outlined with white resin

leaf ½in (1.5cm) long

cylindrical, upright cone

Korean fir cones
Dark purple female seed cones develop in clusters on the top of shoots.

FEATURES

Bark Dark olive-green; develops light, freckle-like pores (lenticels) with age

Abies koreana (Pinaceae)

KOREAN FIR

This broadly conical conifer originates from the volcanic island of Quelpeart, South Korea, where it grows on mountain slopes up to 3,280ft (1,000m). One of the smallest firs, the Korean fir has found its place as a much-revered garden tree in Europe and North America since its introduction into cultivation in 1908. In the latter part of the 20th century, several ornamental dwarf cultivars were developed from this species.

The short, stubby, curved leaves are dark green on the upper surface, with two white bands on the underside. The leaves almost obscure the fawn-colored, pubescent shoots. Male and female cones appear on the same tree. Red-brown male pollen-bearing cones are normally coated in resin and clustered around side shoots. Dark red to purple female seed cones are positioned on the top of shoots.

◗ **Delavay's fir (*Abies delavayi*)** This is a faster growing, larger conifer compared to the Korean fir, and is also distinguished by its larger, barrel-shaped cones.

PROFILE

- Up to 160 ft (50 m)
- Up to 20 ft (6 m)
- Evergreen
- Needlelike
- W. North America

Noble fir cones
Large green cones, ripening to purple-brown, develop from erect, yellow female cones on the top side of shoots.

cylindrical, flat-topped cone, up to 10 in (25 cm) long

green, protruding bract on cone

FEATURES

Bark Smooth silver-gray; develops resin-filled blisters with age

leaf up to 1 ½ in (3.5 cm) long

Leaves Flattened, blunt-tipped, with central groove; with two silver bands on underside

Abies procera (Pinaceae)

NOBLE FIR

The natural range of this large, stately conifer extends down the west coast of North America, from northern Washington state to California's Siskiyou Mountains. This evergreen was discovered and introduced in Europe by Scottish plant collector David Douglas in 1825.

The noble fir has a long, straight trunk. Pale reddish brown shoots are obscured by flattened leaves that part below the shoot and sweep upward at the sides. The leaves have a bluish green upper surface and two narrow bands of stomata (pores) on their underside. When crushed, the leaves emit a pungent, slightly "catty" aroma. Bright crimson male cones are borne along the underside of shoots.

Colorado white fir
Silver-blue foliage and smaller, narrow, cylindrical cones distinguish this tree (*Abies concolor*).

PROFILE

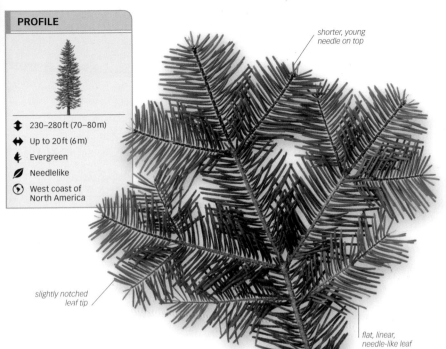

↕ 230–280 ft (70–80 m)

↔ Up to 20 ft (6 m)

🌿 Evergreen

🌿 Needlelike

⊙ West coast of North America

shorter, young needle on top

slightly notched leaf tip

Flat foliage
The flattened leaves are arranged in a single plane on either side of the shoots. They have pale undersides with two green-white bands of stomata (pores).

flat, linear, needle-like leaf

glossy dark green leaf upper surface

FEATURES

Bark Brownish gray turns purple-gray; cracks into square plates in maturity

Female cones Green when young; cylindrical, upright

Abies grandis (Pinaceae)

GRAND FIR

The grand fir is an extremely fast-growing conifer reaching a height of 130 ft (40 m) in 50 years. It is native to the Pacific Northwest and northern California regions, where specimens regularly reach up to 280 ft (80 m) in height, with a trunk diameter of 6 ft (2 m). Since its discovery by Scottish botanist David Douglas in 1830, it has been widely cultivated elsewhere.

The bark on young trees is brownish gray with resin blisters. The tree has a conical crown with whorls of branches, becoming broadly columnar in old age. The foliage emits a fruity, orange fragrance when crushed. Small, ovoid, purple male pollen cones, up to 1/16 in (2 mm) long, are borne on the undersides of shoots. The female seed cones are cylindrical and green, up to 4 in (10 cm) long and 1 ½ in (4 cm) wide, becoming red-brown when mature.

♧ **Alpine fir (*Abies lasiocarpa*)** This tree is distinguished from the grand fir by its smaller stature and distinctive, pale, grayish green, needlelike leaves.

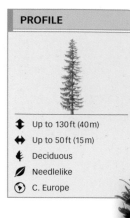

↕ Up to 130 ft (40 m)

↔ Up to 50 ft (15 m)

🌢 Deciduous

🌿 Needlelike

🌐 C. Europe

Larch branchlet
Leaves grow in dense rosettes on side shoots, while appearing singly on main branches. Green, upright seed cones develop from female cones in spring.

scale tip turns slightly inward

blunt, ovoid cone, up to 1 ¾ in (4.5 cm) long

bright green leaf, up to 1 ½ in (4 cm) long

FEATURES

upright, female seed cone

pink-yellow male pollen cone

Cones Male and female cones grow on the same shoot; males on underside

Larix decidua (Pinaceae)

EUROPEAN LARCH

This attractive conifer originates in the Alpine regions of Austria, Switzerland, and Germany, through to the Carpathian mountains of Slovakia and Romania, where it grows up to 8,200 ft (2,500 m) above sea level. It has been widely planted throughout Europe and North America for forestry and ornamental purposes.

Young trees display a smooth, pale gray bark. In older trees, the bark is gray-pink, with vertical fissures and ridges. Well-spaced whorls of upswept branches produce a narrowly conical outline, which is softened by pendulous, straw-colored shoots. The leaves are bright green in spring, turning golden yellow in fall. They are soft at first, but stiffen with age. The female cones appear before the leaves in early spring. The color of mature female seed cones varies from purple-pink to red.

🜄 **Siberian larch (*Larix sibirica*)** Although similar to European larch, this smaller, slower-growing larch can be distinguished by its more conical-shaped, slightly longer cones.

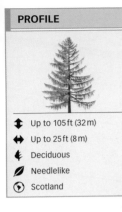

- Up to 105 ft (32 m)
- Up to 25 ft (8 m)
- Deciduous
- Needlelike
- Scotland

cone scale bends outward

egg-shaped cone, up to 2 in (5 cm) long

vibrant green, needlelike leaf in dense rosette

Cones and leaves
The cones are borne upright on shoots next to soft leaves that fade to glaucous blue-green in summer and turn golden brown in fall.

upright female seed cone

Cones Cream to pink or red female seed cone; drooping, yellow male pollen cone

Larix x *marschlinsii* (Pinaceae)

HYBRID LARCH

This broadly conical conifer is also known as Dunkeld larch, after the Scottish estate where it was created in the late 19th century by cross-pollination between the European larch (p.76) and the Japanese larch (p.78). This tree is now more widely planted than either of its parents. It is particularly widespread in forestry plantations because its vigorous growth enables it to reach timber harvesting size quickly. It is also more resistant to disease than either parent.

The bark is reddish brown and scaly in maturity. The shoots are pendulous, orange-brown to pink-buff, and slightly glaucous. On leading shoots, leaves are borne singly and are up to 3 in (8 cm) long. Elsewhere, leaves occur in dense rosettes and are up to 1½ in (4 cm) long. Hybrid larch cones are larger than those of the Japanese larch.

♤ **Tamarack (*Larix laricina*)** This tree is smaller and slower-growing than the hybrid larch. Its needlelike leaves are smaller and thinner, up to 1 in (2.5 cm) long.

seed cone 1¼ in (3 cm) long and wide

seed cone scale turns outward and downward

leaf ¾ in (2 cm) long

leaves borne in rosettes on side shoots

Seed cones and leaves
The Japanese larch seed cones resemble wooden rose blooms, with scales turned outward and downward. The leaves are borne singly on leading shoots and in rosettes on side shoots.

PROFILE

- ↕ Up to 120 ft (35 m)
- ↔ Up to 50 ft (15 m)
- 🍂 Deciduous
- 🌿 Needlelike
- ◉ Japan

FEATURES

Bark Reddish brown; becomes scaly with age

new leaves

Female cones Borne in clusters, pink with cream margins

Seed cones Bun-shaped; green maturing to brown

Larix kaempferi (Pinaceae)

JAPANESE LARCH

Native to Honshu, Japan, this vigorous tree grows at altitudes of up to 9,020 ft (2,750 m) in the wild. It is also grown in and around Japanese temple gardens, where it may be trained as a bonsai. The Japanese larch was introduced in the West by British nurserymen Veitch & Sons in 1861, and has been planted commercially for its strong, durable timber ever since.

It is broader than the European larch (p.76), with long, upswept branches that become horizontal or even slightly descending in older trees. The leaves are flatter and broader as well. They are bright green at first, dulling to gray-green before turning orange in fall. Shoots are reddish purple and conspicuous on leafless branches in winter. Yellow male pollen cones appear in rounded clusters on shoots. Female seed cones emerge in early spring before the leaves.

🌸 **Himalayan larch (*Larix griffithii*)** Unlike the Japanese larch, this tree has long, pendulous, and down-covered shoots. Its large cylindrical cones grow up to 4 in (10 cm) long.

PROFILE

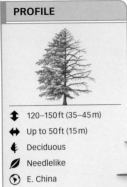

- ↕ 120–150 ft (35–45 m)
- ↔ Up to 50 ft (15 m)
- 🍂 Deciduous
- ⬗ Needlelike
- ⊙ E. China

bright orange-gold
fall leaf

leaf up to 2½ in
(6 cm) long

Vibrant fall foliage
The leaves are borne in dense
clusters on older branches.
They turn orange-gold in
fall before being shed.

FEATURES

bright grass-green
spring leaf

spirally
arranged leaf

Leaves Soft, needlelike;
spirally arranged around
young shoots

Pseudolarix amabilis (Pinaceae)

GOLDEN LARCH

This beautiful, slow-growing conifer was introduced
into western cultivation by plant collector Robert Fortune
in 1852. Despite the name, the golden larch is not a true
larch, having cones that disintegrate on the tree, and
spurs, or side shoots, which lengthen annually.

This tree has gray-brown bark,
smooth at first, becoming deeply
fissured into thick, square plates.
The shoots are pale yellow to
pink-buff. Both male and female
cones are yellow and are borne
at the end of the shoots on the
same tree. Female seed cones
only develop into 2½ in- (6 cm-)
long mature cones after long,
hot summers. Each cone is green,
with triangular scales, maturing
to golden brown before
disintegrating on the tree.

Curving branches
The golden larch has a
broadly conical habit, with
level branches that curve
upward toward the tips

PROFILE

sharply pointed leaf

slight sheen on leaf

- Up to 160 ft (50 m)
- Up to 50 ft (15 m)
- Evergreen
- Needlelike
- N., C., W., and E. Europe

green cone turns brown with age

leaf up to ¾ in (2 cm) long

tapering, curved cone, up to 6 in (15 cm) long

Cones of the spruce
Female cones borne across the top half of the tree become cylindrical and pendulous on stout stems.

FEATURES

Bark Coppery-bronze to pink; matures to dark purple-brown

golden male pollen cone

purple-red female seed cone

Cones Male and female cones borne on the same tree

Picea abies (Pinaceae)

NORWAY SPRUCE

This common conifer, recognized as the traditional Christmas tree throughout most of Europe, has a natural range extending from the Pyrenees to western Russia. It is an important timber-producing tree, used in construction and for making the carcasses of violins and other stringed musical instruments.

It has a symmetrical form, with horizontal branches that gradually become upswept toward the top of the tree. The slightly rough bark develops shallow, round or oval plates that lift away from the surface with age. Male pollen cones cluster at shoot tips, releasing clouds of yellow pollen in spring. When crushed, the rich green leaves emit a citrus-like fragrance.

⌂ Dragon spruce
This tree (*Picea asperata*) resembles the Norway spruce, but its blue-green leaves are more rigid and sharply pointed.

PROFILE

🌲 100–180 ft (30–55 m)

↔ Up to 50 ft (15 m)

🌿 Evergreen

🍃 Needlelike

◉ Caucasus region, N.E. Turkey

Dense foliage
This tree has a conical form with short, blunt-tipped, needlelike leaves.

leaf up to ½ in (1 cm) long

glossy, rigid, blunt-tipped leaves

FEATURES

Bark Smooth, pinkish brown; flakes into small irregular plates after 30 years

patches of sticky resin

Cones Cylindrical, purple; ripening to brown

Picea orientalis (Pinaceae)

ORIENTAL SPRUCE

Also known as the Caucasian spruce, this graceful tree is one of the most attractive and best spruces for cultivation, being tolerant of lime-based soils and better able to withstand drier conditions. Native to the Caucasus region and northeast Turkey, where it grows in mountainous regions up to 7,220 ft (2,200 m), it was introduced into cultivation in 1837. It has been widely planted in parks and gardens ever since, especially in the United States.

The blunt leaves are shiny and dark green on the upper surface, paler on the underside, and uniformly point forward on buff-colored shoots. Both male and female cones appear in separate clusters on the same tree. The male pollen cones turn from bright red to yellow as they mature. The female cones are also red and develop into drooping cones up to 4 in (10 cm) long.

↯ **Likiang spruce (*Picea likiangensis*)** Unlike the oriental spruce, this related tree has flattened leaves, which are bluish green on the upper surface and blue-white on the underside.

Leaves and pendulous cones
The Sitka spruce has blue-green leaves, radially arranged around pale brown shoots. Pale brown hanging cones appear on the shoots in summer.

rigid, sharply pointed leaf, up to 1¼in (3cm) long

thin, papery cone scale

pendulous cone, up to 4in (10cm) long

PROFILE

- ⬍ 230–315ft (70–95m)
- ⬌ Up to 50ft (15m)
- Evergreen
- Needlelike
- North America

Picea sitchensis (Pinaceae)

SITKA SPRUCE

Named after Sitka Island off the Alaskan coast, where it grows in abundance, the Sitka spruce is the largest North American spruce. Valued for its strong, lightweight timber, it was originally used for making aircraft frames, and is now used in the paper industry.

The Sitka spruce has an open, conical habit, with slender, widely spaced, ascending branches. Young trees have a deep purple-brown bark; this lightens on older trees and develops large, curving cracks, which then turn into thin plates of lifting bark. The leaves have two white bands of stomata (pores) on their undersides. Red male and reddish yellow female cones are borne on the same tree in spring.

◔ **White spruce**
Softer, more rounded, lighter-colored leaves and narrow, cylindrical cones distinguish this tree (*Picea glauca*).

PROFILE

↕ 100–160 ft (30–50 m)

↔ Up to 30 ft (10 m)

🗲 Evergreen

◙ Needlelike

⏵ C. Balkans

Spreading foliage
Slender leaves have a dark glossy green upper surface and two broad bands of stomata (pores) on the underside.

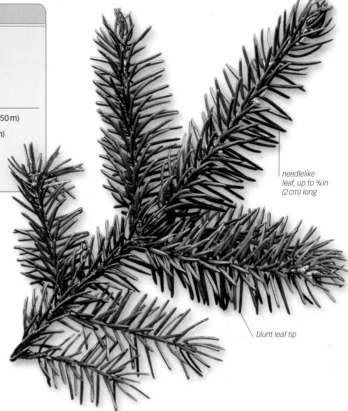

needlelike leaf, up to ¾ in (2 cm) long

blunt leaf tip

FEATURES

Bark Orange-brown to copper; breaks into irregular plates

Picea omorika (Pinaceae)

SERBIAN SPRUCE

Native to the central Balkans, the Serbian spruce faces increasing threat from deforestation, and is now considered vulnerable in the wild. However, it is commonly cultivated in gardens and arboreta as an ornamental tree. This conifer has a slender, spire-like habit, with branches sweeping downward before arching upward at the tips. This form allows the tree to shed snow and so resist damage.

The bark is rarely seen because branching and foliage descends to the ground, obscuring the trunk. Crimson male pollen cones are borne below new shoots, while red female seed cones are confined to the uppermost branches, where they become pendulous in shape. Deep purple-brown and shaped like teardrops, each cone is held on a curved stalk, up to 2½ in (6 cm) long.

◌ **Yezo spruce (*Picea jezoensis*)** Also known as the Hondo spruce, this tree is distinguished by a broader conical form, lighter leaf underside, and longer, red cones.

PROFILE

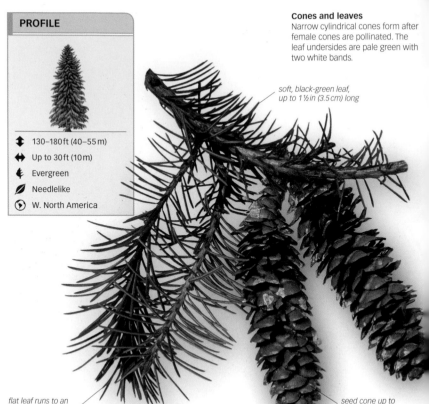

- ↕ 130–180 ft (40–55 m)
- ↔ Up to 30 ft (10 m)
- ♣ Evergreen
- ⬮ Needlelike
- ◉ W. North America

Cones and leaves
Narrow cylindrical cones form after female cones are pollinated. The leaf undersides are pale green with two white bands.

soft, black-green leaf, up to 1½ in (3.5 cm) long

flat leaf runs to an abrupt fine point

seed cone up to 5 in (12 cm) long

FEATURES

Bark Dull pinkish gray color; turns purple-brown with age

Male cones Spherical, purple; up to ¾ in (2 cm) wide

Picea breweriana (Pinaceae)

BREWER'S WEEPING SPRUCE

Possibly the most beautiful of all spruces, this relatively slow-growing, evergreen tree is popular as an ornamental specimen. It is native to the Siskiyou Mountains of northwest California and southwest Oregon. Brewer's weeping spruce is easily recognized by its weeping foliage, which hangs on branchlets like long verdant curtains—sometimes up to 6 ft (2 m) in length.

The purple-brown bark cracks into hard circular flakes as the tree matures. The soft, flat, needlelike leaves are shiny on the upper surface and paler on the underside. They point forward along soft pink-brown shoots, which hang downward like tails. Female seed cones are dark red and clustered on branches near the top of the tree. They become purple as they mature, slightly curving along the length and tapering toward the base. These turn red-gray once the seeds are released.

PROFILE

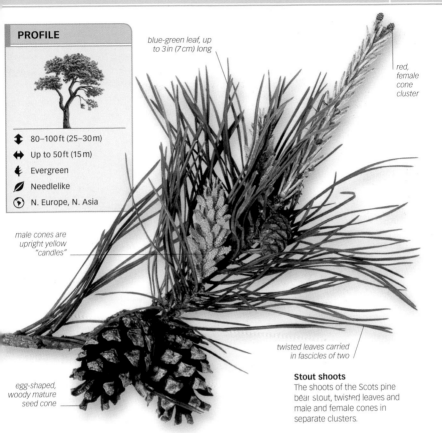

blue-green leaf, up to 3 in (7 cm) long

red, female cone cluster

🔼 80–100 ft (25–30 m)

↔ Up to 50 ft (15 m)

🌿 Evergreen

🌿 Needlelike

⊙ N. Europe, N. Asia

male cones are upright yellow "candles"

egg-shaped, woody mature seed cone

twisted leaves carried in fascicles of two

Stout shoots
The shoots of the Scots pine bear stout, twisted leaves and male and female cones in separate clusters.

FEATURES

Bark Flaking, orange-red in mature trees

raised "pimple" on scale

Seed cones
Green maturing to light brown

Pinus sylvestris (Pinaceae)

SCOTS PINE

This familiar tree is easily recognized by its characteristic red bark. One of the northern hemisphere's most popular trees, it has long been used as a Christmas tree in North America. Its natural range extends from Ireland, across northern Europe and Asia, to the Pacific coast. It can grow on all types of soil, from dry and sandy to wet and peaty. It produces large quantities of seed and is able to colonize new territory quickly. Mature trees develop broadly domed crowns with large snaking branches.

The bark is sometimes red on young trees, although gray or light red-brown is more common. Stiff leaves are set in a papery, orange-brown sheath known as a fascicle. Male and female cones are borne on young shoots in late spring on the same tree.

🔾 **Japanese black pine**
This tree (*Pinus thunbergii*) can be distinguished from the Scots pine by its stout, twisted branches and longer leaves.

PROFILE

- ⬍ 80–100 ft (25–30 m)
- ⬌ Up to 70 ft (20 m)
- ❧ Evergreen
- ∅ Needlelike
- ◉ S. Europe

Paired foliage
The slender, needlelike leaves of this pine are forward-pointing and borne in pairs.

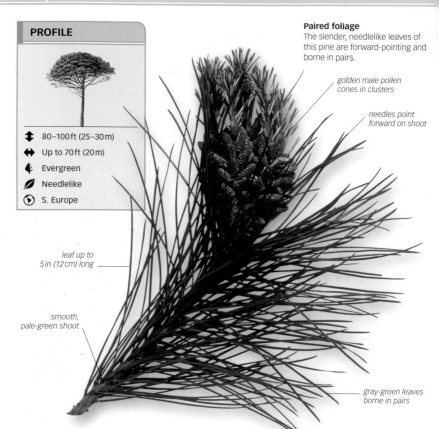

golden male pollen cones in clusters

needles point forward on shoot

leaf up to 5 in (12 cm) long

smooth, pale-green shoot

gray-green leaves borne in pairs

FEATURES

Bark Gray, with orange fissure lines down the trunk

cone up to 4 in (10 cm) long

Seed cones
Spherical, ripen from green to glossy brown

Seeds Up to 100 edible seeds released from each cone

Pinus pinea (Pinaceae)

UMBRELLA PINE

Also known as stone pine, this large conifer has a natural distribution that extends across the Mediterranean region, from Portugal to Turkey. It is widely cultivated elsewhere for its seeds, commonly known as pine nuts.

With its spreading, flat-topped crown, made up of long, horizontal branches and dense foliage, the umbrella pine is a distinctive part of the Mediterranean landscape.

Its trunk is normally short and branches out from relatively low down. The shoots are tipped by chestnut-red buds, fringed with white hairs. The cone is relatively smooth, with a flat base and may weigh up to 14 oz (400 g). After forming, it remains closed for up to three years before opening to release its seeds.

⚲ **Macedonian pine**
Unlike the two-needled umbrella pine, this five-needled tree (*Pinus peuce*) has a narrow conical habit.

PROFILE

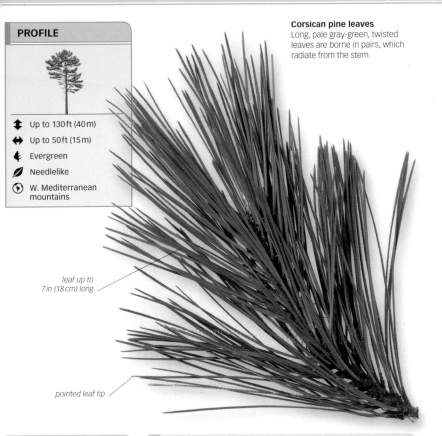

Corsican pine leaves
Long, pale gray-green, twisted leaves are borne in pairs, which radiate from the stem.

↕ Up to 130ft (40m)

↔ Up to 50ft (15m)

🕭 Evergreen

▱ Needlelike

◉ W. Mediterranean mountains

leaf up to 7in (18cm) long

pointed leaf tip

FEATURES

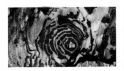

Bark Light gray to pink; heavily fissured from an early age

cone matures to gray-brown

Seed cones Ovoid to conical, up to 3in (8cm) long

Pinus nigra subsp. *laricio* (Pinaceae)

CORSICAN PINE

This large conifer is valued for its strong timber, which is widely used in the construction industry. In recent years, the tree has been prone to attack from red-band needle blight disease. The Corsican pine differs from the "parent" species Austrian pine (*Pinus nigra*) in having a more open crown with fewer and shorter branches, which are level rather than ascending.

Stiff yellow-brown shoots bear sharply pointed buds, often covered in white, crystallized resin. The leaves are carried sparsely on the shoots. Golden yellow pollen cones appear at the base of the shoots. These shed dustlike pollen in late spring and early summer. Dull pink seed cones become slightly curved and woody as they mature.

⌂ **Arolla pine**
This tree (*Pinus cembra*) has a smaller stature and shorter leaves, borne in groups of five.

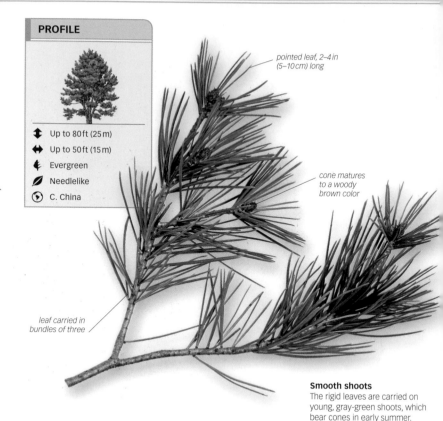

pointed leaf, 2–4 in
(5–10 cm) long

cone matures
to a woody
brown color

leaf carried in
bundles of three

Smooth shoots
The rigid leaves are carried on
young, gray-green shoots, which
bear cones in early summer.

PROFILE

- ↕ Up to 80 ft (25 m)
- ↔ Up to 50 ft (15 m)
- ✦ Evergreen
- ⬤ Needlelike
- ◉ C. China

FEATURES

Bark Gray-green; peels to
reveal a range of colors

Cones Small, squat, ovoid; up
to 3 in (7 cm) long; each tipped
with a spine

Pinus bungeana (Pinaceae)

LACEBARK PINE

This slow-growing, medium-sized conifer is native
to central China and was first seen by outsiders in
1831, growing in a temple garden near Beijing. It was
introduced into western cultivation in 1846. Since then,
it has been planted regularly in gardens and parks. It
has a rounded top, often developing multi-stems
from near the base.

The tree has smooth, silky bark, which is gradually
shed to reveal patches of new bark, rather like the
London plane (p.204). These patches then turn a
succession of colors including cream, pale yellow,
green, red, olive-brown, and purple—quite often a
patchwork of all. The overall effect can be beautiful
and almost reptilian. Male and female cones
are borne in separate clusters on shoots.

♧ **Armand pine (*Pinus armandii*)** The bark of this pine is not
as attractive as the lacebark pine. However, it has longer and more
decorative leaves and cones, both up to 6 in (15 cm) long.

PROFILE

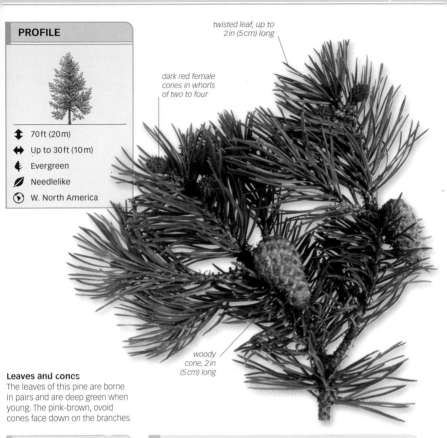

twisted leaf, up to
2 in (5 cm) long

dark red female
cones in whorls
of two to four

woody
cone, 2 in
(5 cm) long

↕ 70 ft (20 m)

↔ Up to 30 ft (10 m)

🕯 Evergreen

⬮ Needlelike

◉ W. North America

Leaves and cones
The leaves of this pine are borne
in pairs and are deep green when
young. The pink-brown, ovoid
cones face down on the branches.

FEATURES

Bark Dark gray, with irregular
chestnut-colored fissures

pollen
cone in
dense
whorl

Male cones Yellow; shed
pollen in spring

Pinus contorta var. *latifolia* (Pinaceae)

LODGEPOLE PINE

This hardy conifer gets its common name from
its traditional use by Native Americans for the central
pole of their tepees or lodges. Native to the mountains
of western North America, the lodgepole pine was
introduced into Europe by
Scottish botanist David Douglas
in 1831. It is able to withstand
exposure, cold, and wet soils
extremely well.

The tree grows straight with
defined whorls of branches,
sometimes as much as 3 ft (1 m)
apart. It is broad and bushy
near the ground, often with a
hockey-stick-like sweep to the
base of the trunk. The leaves
are borne in dense clusters
screening rigid shoots from view,
and turn yellow with age.

⚲ **Weymouth pine**
Unlike the lodgepole pine,
this large, round-headed
tree (*Pinus strobus*) bears
its glaucous green leaves
in groups of five.

— golden new shoot

*bright yellow
male pollen cone*

- 100–130ft (30–40m)
- Up to 50ft (15m)
- Evergreen
- Needlelike
- California

*needlelike
leaf, 4–6in
(10–15cm) long*

Leaves and pollen cones
Pine shoots have dark green leaves,
borne in threes. Male pollen cones,
in dense clusters on shoot tips,
shed pollen in early spring.

FEATURES

— *immature
purple
seed cone*

Seed cones
Borne in clusters
on young shoots

*bulging scales
distort shape
of cone*

Mature cones Initially
red-brown; up to 4in (10cm)
long, roughly oval shape

Pinus radiata (Pinaceae)

MONTEREY PINE

The natural range of this fast-growing conifer is limited to
Monterey Peninsula, California. Since its discovery in 1833,
it has become the most widely planted tree in the world
for timber production. In New Zealand alone, it accounts
for over 60 percent of all conifers, covering more than
2,500 sq miles (4,000 sq km).

In maturity, this tree broadens
out to form a large, domed crown.
The bark is slate-gray, becoming
deeply fissured with age. In
spring, the male pollen cones
shed large amounts of pollen,
while female cones develop into
red-brown seed cones, which age
to dark gray. They usually take
several years to open and release
their seeds. They may stay fixed
to branches for many years even
after seed dispersal.

♤ Big-cone pine
This tree (*Pinus coulteri*)
has large cones weighing
up to 4½lb (2kg) and
stiffer, longer needles
than the Monterey pine.

PROFILE

- 100 ft (30 m)
- Up to 50 ft (15 m)
- Evergreen
- Needlelike
- Canary Islands

Pollen cones
Clusters of small, beige, male pollen cones appear as stout "candles" on branch tips.

individual ripe pollen cone, up to ½ in (1 cm) long

FEATURES

Bark Red-brown, thick, fissured, flaking

Leaves Slender, lax, up to 12 in (30 cm) long; borne on pendulous shoots in clusters of three

Pinus canariensis (Pinaceae)

CANARY ISLAND PINE

Widely planted as an ornamental conifer across warmer regions of Europe, the natural spread of this tree is limited to islands in the Canary archipelago, except Lanzarote and Fuerteventura. It is one of the dominant trees in the Canary Island montane cloud forests, such as those on the western and northern flanks of Mount Teide on Tenerife. Here, the pine's long leaves "comb" low-lying clouds for valuable moisture.

When young the tree is conical in shape. It becomes more widely spreading, with long branches and drooping branchlets, as it matures. Young glaucous leaves turn bright grass-green as they age. The margin of each leaf is finely toothed and feels coarse when run through the fingers. The egg-shaped, cylindrical seed cones, up to 10 in (25 cm) long, with raised stubs on each scale, are rough to the touch.

♤ **Mountain pine (*Pinus mugo*)** This related tree is distinguished by its shrublike sprawling habit, and its short, rigid leaves, up to 1½ in (4 cm) long, which grow in pairs.

PROFILE

pink-purple female seed cone, near tips of new shoots

Soft needles
Although needlelike in form, the leaves of the Bhutan pine are slender and flexible. They are held in fascicles (bundles) of five by a basal sheath.

long leaf, up to 8 in (20 cm) long

- ↕ 100–130 ft (30–40 m)
- ↔ Up to 50 ft (15 m)
- ↟ Evergreen
- ⌀ Needlelike
- ⊙ Himalayas

FEATURES

Male cones Pale yellow; pollen bearing; borne at the base of new shoots

curved cone contains resin

cone up to 12 in (30 cm) long

Seed cones Pendulous, green ripening to pale brown; release seeds in the second year

Pinus wallichiana (Pinaceae)

BHUTAN PINE

Also called the Himalayan white pine, the Bhutan pine is among the most popular pines across the world and is widely planted in parks and gardens. In the Himalayas, this broadly conical tree grows at altitudes of up to 12,500 ft (3,800 m) and reaches 130 ft (40 m) in height.

As with many other pines, it is conical when young but develops a "middle-age spread," with lower branches curving downward from the trunk before sweeping up at the tips. Smooth branches emerge in regularly spaced whorls. The shoots are slate-gray with a purple bloom. The pink-gray bark has some resin blisters, and develops small fissures in maturity. This tree makes good firewood, and is also used to make turpentine.

◊ Korean pine
This tree (*Pinus koraiensis*) has a smaller stature and rough, stiff, needlelike leaves.

PROFILE

- 100–160 ft (30–50 m)
- Up to 50 ft (15 m)
- Evergreen
- Needlelike
- Mexico, W. Central America

slender, spreading foliage

Graceful, weeping foliage
This tree has long, drooping leaves, sometimes bent along their length, held in clusters by a papery sheath attached to the shoots.

blue-green, drooping leaf, up to 6 in (15 cm) long

FEATURES

sticky, white resin cover

Seed cones Long, slender, up to 18 in (45 cm) long, 4 in (10 cm) wide; large scales

Pinus ayacahuite (Pinaceae)

MEXICAN WHITE PINE

Native to Mexico and western Central America, the Mexican white pine grows to more than 130 ft (40 m) in its homeland; elsewhere it is more likely to grow up to 100 ft (30 m). It has been cultivated in Australasia, South Africa, northern India, and temperate regions of Europe.

The tree has pale green shoots, with fine hairs along the length. Young trees often produce a sinuous leading shoot, which develops into long, level branching on old trees. The dark purplish brown bark is smooth at first, coarsely flaking into rough, irregular plates later. The needlelike leaves are finely serrated around the margin. Yellow male and red female cones are borne in separate clusters in early summer. The pendulous seed cones are borne on stout stalks, up to ¾ in (2 cm) long. They are usually covered with sticky, white resin. The scales open to release the seeds inside.

♤ **Holford pine (*Pinus x holfordiana*)** A hybrid derived from the Mexican white pine, the Holford pine is a wider spreading tree with longer needles and less-tapering cones.

FLOWERING TREES

Also known by the scientific term angiosperms, most trees belong to this class. Flowering trees reproduce by producing flowers and have their seeds within a fleshy casing called an ovary that develops into fruit.

Flowering trees can be divided into three subclasses: primitive angiosperms (opposite), monocotyledons (p.113), and eudicotyledons (p.121). They evolved over 125 million years ago, and today there is a huge diversity of flowering plant species.

CLASS	ANGIOSPERMAE
SUBCLASS	3
ORDERS	40
FAMILIES	603
SPECIES	ABOUT 255,000 (c.60,000 TREES)

Flowers are reproductive shoots: these may be large and brightly colored, or small and simple. The basic structure—its male and female parts (p.18) encircled by petals and sepals—is often showy and scented to attract insects, and supplies nectar as an incentive. But two more innovations are equally important in flowering trees. Firstly, they exhibit double fertilization: each pollen grain produced by the male anther supplies the female ovule with two male cells at pollination. One fertilizes the egg to make the embryo, while a second fuses with neighboring tissue to make a seed nutrient store. Secondly, seeds are enclosed in fleshy encasements called fruits. The nutrient store nurtures the embryo; fruit may lure animals to help scatter seeds.

Cockspur hawthorn
This tree produces many flowers, the scent of which is reminiscent of rotting flesh and attracts hungry carrion beetles to pollinate them.

PRIMITIVE ANGIOSPERMS

This group—also known as basal angiosperms—encompasses a handful of unrelated plant groups that were the first flowering plants to evolve. Some of these developed the tree habit and a few—such as magnolias—are notable for their showy flowers.

Flowering plants used to be classified into two subclasses: the monocotyledons (monocots), containing mostly non-woody plants (such as palms, orchids, and grasses) and dicotyledons (dicots), containing all other groups. It is now known that there are several dicot lineages, some of which—loosely called the "primitive angiosperms"—were evolutionary pioneers in reproduction by flowers.

CLASS	ANGIOSPERMAE
GROUP	BASAL ANGIOSPERMS
ORDERS	9
FAMILIES	29
SPECIES	7350 (c.3000 TREES)

BROAD LEAVES

These plants produced some of the first flowers over 130 million years ago, and many produced broader leaf blades for improved photosynthesis (p.9). In the constantly moist warm tropics these trees could stay evergreen, whereas in temperate and seasonally dry regions they turned deciduous. But everywhere the succulent foliage of these first broad-leaved forests attracted browsing animals, such as dinosaurs. These early flowering plants lacked the distasteful waterproofing resins of conifers, so their leaves evolved an alternative cocktail of chemicals—with tannins and even cyanides—specifically to deter herbivores. The descendants of these bitter-leaved basal angiosperms are still with us today, most belonging to a group collectively known as magnoliids: laurels, avocado, nutmeg and of course the magnolias.

SIMPLE FLOWERS

Most magnoliids produce flowers with a straightforward arrangement of parts, and some magnoliid trees have large flowers—such as those of magnolia themselves. Their petals are usually indistinguishable from the sepals (so are called tepals), and typically each flower has many male parts (stamens) and female parts (carpels). Many species are pollinated by beetles—an ancient group of insects, whose prehistoric diversification occurred at a time when the magnoliids were coming to prominence. The fruits of basal angiosperms are usually small and dry.

Magnolia flowers
The bright, large flowers of magnolias, such as these from *Magnolia* x *loebneri* 'Leonard Messer,' are made up of tepals rather than petals, and are particularly attractive to pollinating insects.

PROFILE

- ↕ 60–80ft (18–25m)
- ↔ Up to 30ft (10m)
- ✿ Evergreen
- ⚘ Alternate simple
- ◔ 3.8 class from North Carolina to Florida and east to Texas)

creamy white to lemon-colored sepals

Fragrant flower
The highly fragrant, creamy white flowers of the bull bay carry the scent of citronella.

flower borne singly at end of shoot

flower opens in a saucer shape, up to 12in (30cm) wide

leaf glossy green on upper surface

FEATURES

light in a fleshy shaped seed

Fruit Red-brown, cone-like structure with seeds

Magnolia grandiflora (Magnoliaceae)

BULL BAY

Also known as loblolly magnolia, this is indigenous to the coastal US. In warm, for it develops into a freestanding tree with a few short stems, and is a common sight in squares and in gardens. In cooler often grown as a high wall shrub, where the combination of glossy, dark green foliage and creamy white flowers is dramatic.

The gray-brown bark is smooth at first, cracking into irregular small plates as it ages. The leaf undersides are sometimes covered in copper-colored hairs. Flowers, borne through summer and early fall, provide a late source of nectar for insects, are followed by erect seed displaying red seeds.

PRIMITIVE ANGIOSPERMS

This group—also known as basal angiosperms—encompasses a handful of unrelated plant groups that were the first flowering plants to evolve. Some of these developed the tree habit and a few—such as magnolias—are notable for their showy flowers.

Flowering plants used to be classified into two subclasses: the monocotyledons (monocots), containing mostly non-woody plants (such as palms, orchids, and grasses) and dicotyledons (dicots), containing all other groups. It is now known that there are several dicot lineages, some of which—loosely called the "primitive angiosperms"—were evolutionary pioneers in reproduction by flowers.

CLASS	ANGIOSPERMAE
GROUP	BASAL ANGIOSPERMS
ORDERS	9
FAMILIES	29
SPECIES	7350 (c.3000 TREES)

BROAD LEAVES

These plants produced some of the first flowers over 130 million years ago, and many produced broader leaf blades for improved photosynthesis (p.9). In the constantly moist warm tropics these trees could stay evergreen, whereas in temperate and seasonally dry regions they turned deciduous. But everywhere the succulent foliage of these first broad-leaved forests attracted browsing animals, such as dinosaurs. These early flowering plants lacked the distasteful waterproofing resins of conifers, so their leaves evolved an alternative cocktail of chemicals—with tannins and even cyanides—specifically to deter herbivores. The descendants of these bitter-leaved basal angiosperms are still with us today, most belonging to a group collectively known as magnoliids: laurels, avocado, nutmeg and of course the magnolias.

SIMPLE FLOWERS

Most magnoliids produce flowers with a straightforward arrangement of parts, and some magnoliid trees have large flowers—such as those of magnolia themselves. Their petals are usually indistinguishable from the sepals (so are called tepals), and typically each flower has many male parts (stamens) and female parts (carpels). Many species are pollinated by beetles—an ancient group of insects, whose prehistoric diversification occurred at a time when the magnoliids were coming to prominence. The fruits of basal angiosperms are usually small and dry.

Magnolia flowers
The bright, large flowers of magnolias, such as these from *Magnolia* x *loebneri* 'Leonard Messer,' are made up of tepals rather than petals, and are particularly attractive to pollinating insects.

PROFILE

- ↕ 60–80 ft (18–25 m)
- ↔ Up to 30 ft (10 m)
- 🌿 Evergreen
- 🍃 Alternate simple
- ◉ S.E. coast (from North Carolina to Florida and east to S.E. Texas)

creamy white to lemon-colored sepals

Fragrant flower
The highly fragrant, creamy white flowers of the bull bay carry the scent of citronella.

flower borne singly at end of shoot

flower opens in a saucer shape, up to 12 in (30 cm) wide

leaf glossy green on upper surface

FEATURES

bright red, kidney-shaped seed

Fruit Red-brown, cone-like structure with seeds

Magnolia grandiflora (Magnoliaceae)

BULL BAY

Also known as loblolly magnolia, bull bay is indigenous to the coastal US. In warm, temperate areas, it develops into a freestanding tree with a broad canopy and short stems, and is a common sight in streets, town squares, parks, and gardens. In cooler regions, it is more often grown as a large wall-shrub, where the combination of glossy, dark green foliage and creamy white flowers is dramatic.

The gray-brown bark is smooth at first, cracking into irregular small plates as it ages. The leaf undersides are sometimes covered in copper-colored hairs. Flowers, borne through summer and early fall, provide a late source of nectar for insects, and are followed by erect seed pods displaying red seeds.

◔ **Cucumber magnolia**
This deciduous tree (*Magnolia acuminata*) has ovate leaves, and green and yellow flowers.

PROFILE

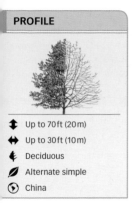

⬍ Up to 70ft (20m)

⬌ Up to 30ft (10m)

🍂 Deciduous

🌿 Alternate simple

◉ China

pale pink tepal

color fades as flower ages

folded leaf bud

tepal splays outward when old

Showy bloom
The flowers of Dawson's magnolia bloom in early spring and may grow up to 10in (25cm) wide. Each flower may have up to 12 tepals.

FEATURES

Leaves Obovate to elliptic, matte with bright green upper surface; 3–6in (7.5–15cm) long

Magnolia dawsoniana (Magnoliaceae)

DAWSON'S MAGNOLIA

Native to Sichuan and Yunnan provinces of China, this broadly conical tree grows in mountainous regions at an altitude of around 8,200ft (2,500m). Discovered in 1869 by the French missionary Père Jean Pierre Armand David, it was introduced into western cultivation when English botanist Ernest H. Wilson sent its seeds to the Arnold Arboretum in Jamaica Plain, Massachusetts, in 1908.

A hardy tree, Dawson's magnolia is able to withstand several degrees of frost. The bark is gray-brown and smooth at first, but on older trees develops some shallow cracks and fissures. Leaves are broadest at the middle and taper toward each end, with a blue-silvery green underside. Bright pink buds bloom to produce masses of slightly fragrant, cup-shaped, pale pink flowers. The fruit is an erect, cylindrical pod up to 6in (15cm) long and contains red seeds.

♧ **Lily tree (*Magnolia denudata*)** Unlike the large, pink flowers of the Dawson's magnolia, this tree has smaller, pure white flowers, which bloom in late spring.

PROFILE

↕ 25–40 ft (8–12 m)

↔ Up to 30 ft (10 m)

🍂 Deciduous

🌿 Alternate simple

◉ Hybrid, originated in Europe

Early bloom
Large, goblet-shaped flowers start to open in early spring, often before the leaves emerge.

flower up to 10 in (25 cm) wide

pink-purple staining at base of tepal

FEATURES

leaf up to 8 in (20 cm) long

Leaves Smooth, dark green on the upper surface; paler and finely hairy on the underside

cylindrical cluster encases seeds

Fruit Irregular cylindrical cluster, ripening from green to pink-red

Magnolia x *soulangeana* (Magnoliaceae)

SAUCER MAGNOLIA

This magnolia was the first to be hybridized and does not occur in the wild. A cross between *Magnolia denudata* and *M. liliiflora*, it was first raised near Paris in the 1820s by Etienne Soulange-Bodin, a French army officer. Today, it is the world's most widely planted garden magnolia, with many cultivated forms such as 'Alba Superba,' 'Brozzonii,' 'Lennei,' and 'Nigra.'

This tree tends to have multiple stems with low branches. The bark is smooth and gray, while flowers vary in color, from creamy white to pink or purple-pink. Leaves are elliptic to obovate, tapering to a pointed tip. Seeds form in clusters from late summer to early fall.

♤ **Loebner's magnolia**
A hybrid of *M. stellata* and *M. kobus,* this tree (*Magnolia* x *loebneri*) flowers later in spring than both parents.

PROFILE

⬍ Up to 70 ft (20 m)

⬌ Up to 30 ft (10 m)

🍂 Deciduous

🌿 Alternate simple

◉ China

Large flowers
The flowers of Sargent's magnolia are among the largest of any magnolia, growing up to 12 in (30 cm) wide.

rose-pink inner surface of tepal

tepal pale pink inside

FEATURES

obovate leaf

Leaves Thick and leathery; up to 8 in (20 cm) long and 4 in (10 cm) wide

Fruit Green ripening to pink; open in summer to reveal glossy, bright red seeds

Magnolia sargentiana (Magnoliaceae)

SARGENT'S MAGNOLIA

This broadly columnar magnolia is native to the provinces of Yunnan and Sichuan in China, from where it was introduced into western cultivation in 1908 by Ernest H. Wilson. It is named after Charles Sprague Sargent, who was director of the Arnold Arboretum in Jamaica Plain, Massachusetts, at the time of its introduction. As a result of deforestation and habitat destruction, this tree is now considered endangered in the wild.

The gray-brown bark is smooth at first but develops shallow cracks and fissures with age. The leaves are dark green with a slight sheen on the upper surface, and have a pale green underside. The fragrant flowers are rose pink on the outside and pale pink on the inner surface, resembling giant water lilies. They are produced in early spring before the leaves unfurl. The fruit is a rigid, cylindrical pod up to 6 in (15 cm) long.

◌ **Umbrella tree (*Magnolia tripetala*)** With thin, lax, large, obovate to elliptic leaves, this tree is also distinguished by its fragrant, cream-colored flowers in early summer.

PROFILE

- Up to 12 ft (4 m)
- Up to 15 ft (5 m)
- Deciduous
- Alternate simple
- Japan

Showy, white blossoms
Flower buds open in spring to reveal starlike flowers made up of up to 20 narrow, white or pink-blushed tepals.

narrow, white tepal up to 2 in (5 cm) long

FEATURES

Bark Gray, smooth with a slight sheen on trunk; brown on branches

paler underside of leaf

deep green upper surface of leaf

Leaves Green, obovate, and narrow; up to 4 in (10 cm) long, 2 in (5 cm) wide

Magnolia stellata (Magnoliaceae)

STAR MAGNOLIA

This slow-growing, broadly spreading, small magnolia is native to the mountains of Nagoya, Japan, from where it was introduced into western cultivation in the 1860s. Some suggest that it is a form of *Magnolia tomentosa* or *Magnolia kobus,* and not a true species in its own right; however it does have visual differences to both of these species. The star magnolia is ideal for planting in small gardens and usually flowers from a young age.

This magnolia's chestnut-brown branches are long, thin, and twiglike. It has few large, heavy branches, so the overall effect is more shrublike than treelike. Conical flower buds, evident throughout winter, are covered in silver-gray down. They open gradually in spring, before the leaves appear.

↻ Northern Japanese magnolia
This tree (*Magnolia kobus*) is larger, with obovate leaves and creamy white tepals, up to 3½ in (9 cm) long.

Conspicuous flower
Large, fragrant flowers open singly from ovoid buds and vary from deep pink to pale pinkish white.

flower up to 12 in (30 cm) wide

spreading, pinkish white outer tepal

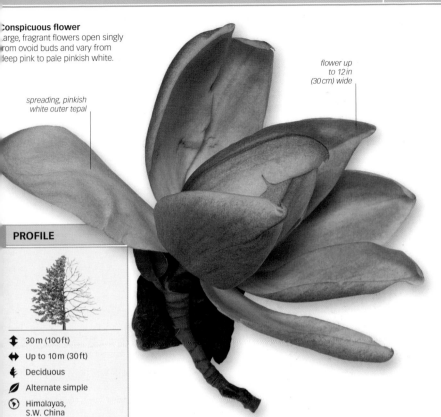

PROFILE

↕ 30 m (100 ft)

↔ Up to 10 m (30 ft)

♣ Deciduous

⬭ Alternate simple

◉ Himalayas, S.W. China

FEATURES

leaf ends in sharp point

Leaf Gray-green, with slight sheen on upper surface

Buds Large, ovoid, covered in short hairs

Magnolia campbellii (Magnoliaceae)

PINK TULIP TREE

Also known as the giant Himalayan pink tulip tree or Campbell's magnolia, this majestic, broadly conical tree grows at heights of up to 9,840 ft (3,000 m) above sea level in its natural range. It is one of the largest magnolias in cultivation, and since the 1860s has been widely planted in parks, large gardens, and arboreta.

This tree has a smooth, mid-gray bark, which it maintains into old age. Obovate leaves grow up to 12 in (30 cm) long, and are pale green, occasionally faintly hairy, on the underside. The flower buds stand out dramatically on bare branches in late winter. The flowers are goblet-shaped at first, opening to a lax cup and saucer shape and have tepals rather than petals and sepals. They open before the leaves and are prone to late frost damage. The tepals fall gradually, and may remain on the ground for some time.

◌ **Veitch's hybrid magnolia (***Magnolia* x *veitchii***)** This smaller related species, up to 52 ft (16 m) tall, is easily distinguished by its compact habit and more goblet-shaped flowers, up to 8 in (20 cm) wide.

PROFILE

↕ Up to 92 ft (28 m)

↔ Up to 50 ft (15 m)

🍂 Deciduous

🌿 Alternate simple

◉ C. and W. China

unripe green fruit, up to 3 in (7.5 cm) long

leaf up to 7 in (18 cm) long

Fruit and leaves
The fruit is a rigid pod, which ripens in summer to reveal bright red seeds. The leaves are obovate with a wedge-shaped base.

FEATURES

Bark Gray; remains smooth in maturity

Flowers Rose pink, fading to light pink with lax tepals in maturity; up to 8 in (20 cm) wide

Magnolia sprengeri (Magnoliaceae)

SPRENGER'S MAGNOLIA

This broadly conical, medium to large-sized magnolia is native to central and western China, from where it was introduced into western cultivation by plant collector Ernest H. Wilson in 1900. This tree is named after the German botanist Carl Ludwig Sprenger. It is similar to the pink tulip tree (p.101), although defining features such as overall size, spread, leaves, and flowers are smaller in Sprenger's magnolia.

The leaves are bright green on the upper surface and pale green on the underside. The ovoid flower buds, up to 3 in (7.5 cm) long, are covered with silver-gray down. They remain on the tree throughout late winter, revealing goblet-shaped, fragrant flowers in mid-spring. On maturity, they open out to form water-lily-like flowers.

◊ **Japanese big-leaf magnolia (*Magnolia obovata*)** This tree has very large, obovate leaves, up to 20 in (50 cm) long and 10 in (25 cm) wide, borne in whorls at the ends of shoots.

PROFILE

- ↕ Up to 25 ft (8 m)
- ↔ Up to 15 ft (5 m)
- 🍂 Deciduous
- 🌿 Alternate simple
- ◑ W. China

flower up to
4 in (10cm) wide

leaf pale
green on
underside

Drooping flowers
The flowers hang down from
arching stems in late spring, and
are seen from beneath the tree.

FEATURES

Fruit Green, cylindrical seed
pod; ripening to purple-pink

Magnolia wilsonii (Magnoliaceae)

WILSON'S MAGNOLIA

This broadly spreading shrublike tree has become
a popular garden plant in recent years. It is named after
English plant collector Ernest H. Wilson, who discovered
the species in western China in 1908.

The bark is green with a bumpy texture. Leaves are
narrowly elliptic, running to a pointed tip. They are up
to 8 in (20cm) long and 4 in (10cm) wide, with a silk-green
upper surface, and a thin, white, feltlike covering on
the underside. The flower buds are covered in dense,
soft gray hairs and are visible from late winter. Fragrant,
cup-shaped, white flowers with crimson-purple stamens
emerge after the leaves appear. Over time, the tepals
spread and the flower becomes saucer-shaped. A green
aggregate fruit, up to 3 in (7.5cm) long, develops after
flowering. It ripens to purple-pink and opens to reveal
many bright orange-red, glossy seeds inside.

◊ **Mulan magnolia (*Magnolia liliiflora*)** With its erect, slender,
tulip-shaped, purple, or purple-flushed flowers, this related species
is unlikely to be mistaken for Wilson's magnolia.

Multi-petaled flowers
Blooming in spring, the sweet michelia's cup-shaped, multi-petaled flowers are pale lemon to white in color.

orange stamens

PROFILE

- Up to 70 ft (20 m)
- Up to 50 ft (15 m)
- Evergreen
- Alternate simple
- China, Himalayas

Michelia doltsopa (Magnoliaceae)

SWEET MICHELIA

This broadly spreading tree is one of the most attractive of all evergreens. It is related to magnolia and was discovered in Yunnan, western China, in 1918 by the Scottish plant collector George Forrest. He sent seeds of the sweet michelia to Caerhays castle in Cornwall, England. These were propagated, producing five seedlings, which were planted in the castle grounds. They flowered for the first time in 1933.

The glossy, dark green, leathery leaves are up to 8 in (20 cm) long, and covered with a fine coating of rust-colored hairs on the underside. The flower buds form in fall and are like erect magnolia buds, but covered in fine cinnamon-colored hairs rather than the silver hairs typical of magnolia. The buds open in spring to reveal cup-shaped flowers. Each flower is strongly scented, almost pungently so in an enclosed space or on hot spring days.

♧ **Chinese evergreen magnolia (*Magnolia delavayi*)** This tree's leaves are almost twice the size of those of sweet michelia. It also bears flowers that have larger, lax tepals.

↕ Up to 160 ft (50 m)

↔ Up to 70 ft (20 m)

🍂 Deciduous

🌿 Alternate simple

⊙ E. North America

Tulip tree flower
Upright, tulip-shaped flowers, 2½ in (6 cm) wide, are produced in summer. Each has six petals and clusters of stamens inside.

light green to orange-yellow petal

orange-yellow stamen

Bark Smooth, gray-brown; becomes vertically and regularly fissured with age

cone-shaped fruit

Leaves Bright green; have four pointed lobes with notched tips

Liriodendron tulipifera (Magnoliaceae)

TULIP TREE

The distinctive tulip tree is native to eastern North America, from Ontario in the north to Florida in the south. Fossil records show it grew in Europe prior to the last ice age. It was introduced into cultivation in England in the mid-17th century. Since then, this large, long-trunked tree has been widely grown in parks, gardens, and arboreta elsewhere.

The leaves are an unusual shape with four lobes and a cut-off indented leaf tip. They are up to 6 in (15 cm) long and 4 in (10 cm) wide, and pale green on the underside. In fall they turn butter-yellow before being shed. The fruit is cone-shaped and contains seeds with papery wings.

Broad and columnar
The tulip tree develops a straight trunk on maturity, with few low branches, topped by a broad canopy.

PROFILE

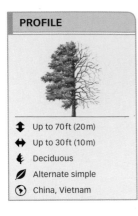

- ⬍ Up to 70 ft (20 m)
- ⬌ Up to 30 ft (10 m)
- 🍂 Deciduous
- ✿ Alternate simple
- 🕑 China, Vietnam

leaf deeply indented between lobes

leaf almost straight at the top

leaf up to 5 in (12 cm) long

Lobed leaves
Leaves run to a lobe on either side, with another lobe midway between the leaf tip and the leaf stalk on each side.

FEATURES

Bark Slate gray, smooth; fissures vertically in maturity

Flowers Pale yellow-green outside, orange-yellow banding inside; has orange stamens

Liriodendron chinense (Magnoliaceae)

CHINESE TULIP TREE

Native to China and Vietnam, the Chinese tulip tree grows up to around 6,560 ft (2,000 m) above sea level. It was introduced into western cultivation by English plant collector Ernest H. Wilson in 1901. Some of the first introductions of this species are still growing in the Royal Botanic Gardens at Kew in London, England.

The leaves are dark green on the upper surface, pale blue-green beneath. Between the lobes, the leaves are more deeply cut than on the tulip tree (p.105), giving the impression that the leaf has a waist. The leaves turn a rich golden yellow in fall. Tulip-like flowers appear singly at the ends of shoots in early to midsummer. The fruit is a pale brown, conical cluster.

Woodland tree
This stately, medium-sized, deciduous tree usually grows in mixed woodland in its natural range.

finely toothed leaf margin

pale green leaf underside, with a lighter midrib

rounded berry, ½ in (1 cm) wide

PROFILE

20–60 ft (6–18 m)

Up to 30 ft (10 m)

Evergreen

Alternate simple

Mediterranean

Foliage and fruit
The leaves are elliptic to lanceolate in shape. Berries form on female trees and turn from green to glossy black when ripe.

berry ripens from green to glossy black

FEATURES

flower ½ in (1 cm) wide

Male flowers Yellow-green with yellow anthers

Female flowers Smaller than male; borne on separate trees

Laurus nobilis (Lauraceae)

BAY LAUREL

This conical and densely leaved tree is found across the Mediterranean. It is widely planted elsewhere for ornamental and culinary purposes, as well as for hedging and screening. The Greeks and Romans used it as a symbol of victory, weaving it into crowns to be worn by champions. They also gave acclaimed poets wreaths made of fruiting sprays, hence the term "poet laureate."

Bay laurel is a multi-stemmed tree, with bark that remains smooth even in old age. The elliptic to lanceolate leaves are dark green and leathery, and give off a rich fruity aroma when crushed. Male and female flowers are borne in leaf axils on separate trees in late winter, followed by berries on female trees.

◊ **Spice bush (*Lindera benzoin*)** Unlike the evergreen bay laurel, spice bush has deciduous leaves, which turn bright yellow in fall, and, on female trees, bright red berries.

PROFILE

- Up to 50 ft (15 m)
- Up to 25 ft (8 m)
- Evergreen
- Alternate simple
- Azores, Madeira, Canary Islands, Morocco

Aromatic leaves and flowers
The elliptic to lanceolate leaves of the Canary Island laurel are very aromatic when crushed. The greenish yellow flowers are also slightly fragrant.

individual flower up to ½ in (1 cm) wide

FEATURES

Bark Dark gray; remains smooth even in maturity

Laurus azorica (Lauraceae)

CANARY ISLAND LAUREL

This medium-sized, broadly columnar tree is an important component of the Laurisilva forests on the northern and western slopes of the Azores, Madeira, and Canary Islands. However, it is now threatened by habitat loss in parts of the Canary Islands. It is often cultivated in parks and gardens in warmer regions of Europe. Like its relative the bay laurel (p.107), this tree's foliage emits a pleasant aroma when crushed.

The thick, leathery, dark green leaves feel rigid to the touch. Paler green and hairy on the underside, they display a wavy-edged margin. The flowers are borne in clusters in the axils of the previous year's leaves. These are followed by the fruit—an ovoid berry, up to ½ in (1.5 cm) long. Green at first, the berry ripens to glossy black.

♧ **Japanese spice tree (*Lindera obtusiloba*)** This tree has ovate to obovate, deciduous leaves, with three large, pointed lobes. They turn bright yellow in fall before being shed.

Bright green leaves
The Chilean laurel bears glossy, bright green leaves, which emit a strong aroma when crushed.

leaf up to 4 in (10 cm) long, 1 ½ in (4 cm) wide

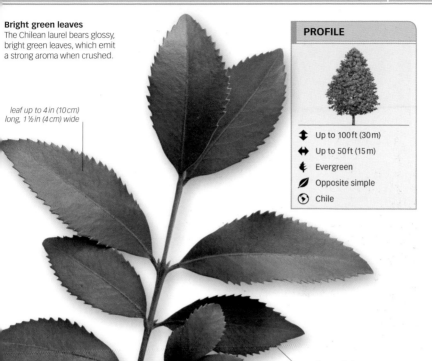

gently serrated leaf margin

PROFILE

↕ Up to 100 ft (30 m)

↔ Up to 50 ft (15 m)

🌿 Evergreen

🍃 Opposite simple

◉ Chile

FEATURES

Bark Pale yellow-gray, smooth; darkens with age

Laurelia sempervirens (Monimiaceae)

CHILEAN LAUREL

Also known as Chilean sassafras, this large, fast-growing, handsome, evergreen tree has long been favored for its attractive timber, which is strong, durable, and highly scented. This has led to the tree becoming threatened in native Chile due to extensive felling. However, it has been introduced as an ornamental tree in the United States and parts of Europe, including Spain and southwest England. It is not hardy enough to grow in regions much farther north than this.

The yellow-gray bark is smooth at first, darkening and cracking into circular plates with age. Its evergreen leaves are glossy and leathery to the touch. Arranged in opposite pairs along the shoot, they are oblong in shape, narrow at the base and tapering to a blunt tip. They are gently serrated around the leaf margin. Clusters of small, slightly fragrant, mustard-yellow flowers appear in the leaf axils in late spring.

PROFILE

- 22–50 ft (7–15 m)
- Up to 22 ft (7 m)
- Evergreen
- Alternate simple
- S. Mexico to Chile and Argentina

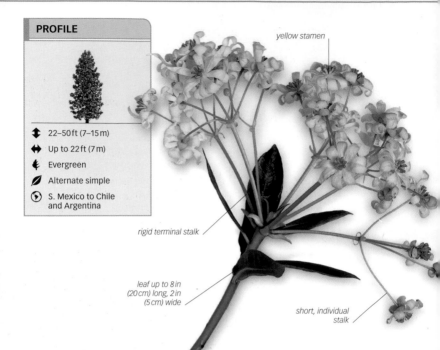

yellow stamen

rigid terminal stalk

leaf up to 8 in (20 cm) long, 2 in (5 cm) wide

short, individual stalk

Fragrant flowers
Jasmine-scented flowers are borne on their own short, lax stalks, but are bunched in a cluster.

FEATURES

unripe fruit

whitish, waxy underside

Leaves Bright green, with a slight sheen on upper surface; glaucous white underside

Drimys winteri (Winteraceae)

WINTER'S BARK

With a natural habitat stretching from southern Mexico to Chile and Argentina, this handsome tree takes its name from Captain William Winter, who sailed with Sir Frances Drake on his famous voyage of discovery in 1577–80. Winter used the bark, high in vitamin C, to treat sailors who were suffering from scurvy. It was first cultivated as an ornamental tree in Europe in 1827.

The bark is relatively soft, brown-gray, and smooth even in old age. The leathery leaves are large, up to 8 in (20 cm) long and 2 in (5 cm) wide. They are widest in the middle and taper to a point at each end. When crushed, they give off a distinctive peppery aroma, as does the bark. The leaves are used in Argentina and Chile as a substitute for pepper. The tree bears ivory-white flowers in late spring, and in late summer and fall they develop into green fruit, which turn blue-black when ripe.

Mountain pepper (*Drimys lanceolata*) Apart from its narrower, smaller leaves, this tree can also be distinguished from the Winter's bark by its purple-red shoots.

leaf up to 18 in (45 cm)
long, 6 in (15 cm) wide

pale green
leaf underside

PROFILE

⬍ Up to 70 ft (20 m)

⬌ Up to 30 ft (10 m)

🌿 Evergreen

🍃 Alternate simple

◐ Mexico, West Indies

Deeply veined leaves
The avocado tree bears elliptic
leaves that display pronounced
veining on their leathery surface.

glossy, dark green
upper surface of leaf

FEATURES

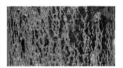

Bark Dark gray, smooth;
becomes vertically fissured

fruit 4–6 in
(10–15 cm) long

oval seed

Fruit Pear-shaped; leathery,
glossy, green to purple-brown
outer skin

Persea americana (Lauraceae)

AVOCADO

Perhaps one of the best known of all tropical fruit
trees, this medium-sized evergreen is native to Mexico
and the West Indies. However, the avocado tree has
been extensively cultivated throughout the tropics for
its flavorsome and highly nutritious fruit, which is rich
in minerals and oils.

The tree has a smooth bark
that becomes fissured with age.
Small, star-shaped flowers are
borne in leaf axils on the ends of
erect, open-branched racemes.
Individual flowers are fragrant,
creamy yellow to white, and up
to ½ in (1 cm) wide. The fruit
occasionally displays mottled
skin, especially on purple-skinned
avocado varieties. Its edible
inner flesh is light green and
surrounds a single seed.

Short-trunked tree
Avocado is a fast-growing,
broadly spreading tree.
It has only a short trunk
before it begins to branch.

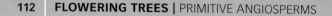

ovate leaf, up to
5 in (13 cm) long

waxy leaf surface

PROFILE

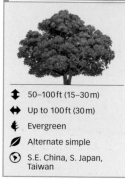

↕ 50–100 ft (15–30 m)

↔ Up to 100 ft (30 m)

🍃 Evergreen

🌿 Alternate simple

⊙ S.E. China, S. Japan, Taiwan

Camphor leaves
In maturity, the leaves are waxy and deep glossy green. They are marked by three distinct veins.

FEATURES

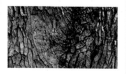

Bark Pale and smooth when young, rough and vertically fissured in maturity

Cinnamomum camphora (Lauraceae)

CAMPHOR TREE

Camphor is a large, broadly spreading evergreen tree, which develops a distinctive rounded canopy of dense, dark leaves in maturity. It is native to southeast China, Taiwan, and southern Japan. Outside these areas this tree is widely cultivated both for its oil and as an ornamental shade tree in parks, gardens, and streets. Camphor also refers to the white, crystalline substance obtained from the wood, bark, and leaves of the tree. It has been used for centuries as a component of incense, an insect repellent, a culinary spice, and a medicine. Its wood is also sought after for its distinctive red and yellow stripes.

The leaves are coppery pink when young, maturing to bright green, and then becoming deep glossy green. When crushed, they emit the fragrance of camphor. Small, white to creamy green flowers are produced in spring, borne on erect racemes in the leaf axils. They are followed by clusters of black, berrylike, rounded fruit, up to 1/2 in (1 cm) wide.

MONOCOTYLEDONS

The non-woody monocotyledons (monocots) generally have elongated strap-like leaves with parallel veins. They have diversified as ground-hugging plants, such as grasses. But several families have evolved trees: most notably the palms.

Most monocotyledon species are low fleshy (as opposed to woody) plants and include many familiar plants: grasses, orchids, and lilies. They get their name from the fact that they have a single pre-packed seed leaf (cotyledon) inside the seed before it germinates.

Monocots are generally incapable of producing woody tissue because their transport vessels are scattered through the stem instead of arranged in wood-forming rings (p.9). Yet the tree habit (p.19) has evolved independently in a number of different monocot groups. Yuccas and cordylines produce leafy branches from a central trunk. Giant "herbs," such as bananas and the traveler's palm, *Ravenala madagascariensis,* strictly have no trunk at all and instead are supported by massive leaf stalks growing from the plant base. But most monocot trees belong to a single family: the palms.

CLASS	ANGIOSPERMAE
SUBCLASS	MONOCOTYLEDONEAE
ORDERS	11
FAMILIES	81
SPECIES	58,000 (c. 3,000 TREES)

TROPICAL GIANTS
The unbranched trunk of a palm is strengthened with fibers reinforced with silica, enabling many to grow to considerable heights. The evergreen fanlike leaves show the familiar compound (palm-leaf) structure, but there is considerable variation and in many species the leaves can grow into giant proportions.

Individually, palm flowers are usually small and inconspicuous, but they may be produced in larger inflorescences (sprays). Most palm species are pollinated by insects, but some are wind-pollinated. Most palms—such as the date palm (p.116) and coconut (p.119)—produce fruits called drupes: here the inner layer of the fruit becomes hardened to form a stone, or shell, which—in turn—encases the true seed. Coconut palms produce large fruits that can float on water, enabling them to colonize oceanic islands. Another member—the Seychelles palm, *Lodoicea maldivica*—produces the world's largest seed in the largest fruit, which weighs more than 66 lb (30 kg) and is too heavy to float anywhere. Palms reach their greatest diversity in the tropics and, although a few have managed to adapt to temperate climates, most only succeed in warm moist forests.

European fan palm
The leaf is typical of the palm family: long with parallel veins, and shaped like a fan. This palm is one of the few that have adapted to temperate climes.

- Up to 40 ft (12 m)
- Up to 10 ft (3 m)
- Evergreen
- Palm
- C. China

Branched flower cluster
The small, fragrant flowers are clustered together in large, bulbous drooping panicles in summer.

stout flower stalk emerges from among leaves

fruit up to ½ in (1.5 cm) wide

creamy yellow male flowers

FEATURES

Bark Covered with brown-gray, fibrous hairs; marked with discarded leaf scars

Leaves Stiff, leathery, dark green, up to 4 ft (1.2 m) wide

Trachycarpus fortunei (Arecaceae)

CHUSAN PALM

This exotic species is also known as windmill palm. Surprisingly, it is not indigenous to the tropics, but originates from the mountains of central China. It is a hardy species and can also grow in warmer regions of Europe and North America. It was introduced into Europe in 1830 by the German physician Philipp Franz von Siebold.

The fan-shaped leaves are divided almost to the base into about 40 linear, pointed strips. The leaves are attached to the slender trunk by a stiff, flattened leaf-stalk, up to 3 ft (1 m) long. Dead leaves may remain on the trunk for years, creating a fringe-like appearance below the new leaves.

Slender trunk
This palm has a small head of clustered leaves at the top of a single trunk, which is narrower at the base.

Dense palm leaves
The European fan palm appears as a dense, semicircular mass of fan-shaped, stiff, gray-green leaves, borne on a short trunk.

stiff, pointed leaf segment, up to 18 in (45 cm) long

PROFILE

↕ Up to 8 ft (2.5 m)

↔ Up to 10 ft (3 m)

🌿 Evergreen

🍂 Whorled compound

📍 Mediterranean region

FEATURES

Bark Brown, fibrous trunk; bears thick leaf stalks

Chamaerops humilis (Arecaceae)

EUROPEAN FAN PALM

This tree is the only member of the palm family native to mainland Europe. In the wild, it grows on coastal mountainsides in Spain, Gibraltar, the Balearics, Sardinia, Italy and Sicily, as well as Algeria and Morocco. It was first cultivated outside this region, in areas such as southwestern Britain, in 1731.

The leaves, up to 36 in (90 cm) wide, are divided near the base into segments or strips. They are attached to the trunk by stalks from the point where the strips converge. The stalks, normally around 12 in (30 cm) long, are covered with sharp, forward-pointing spines. Small, yellow flowers emerge in stiff, upright panicles, up to 6 in (15 cm) long. The fruit, borne in clusters around the leaf base, resemble shiny green olives.

Small, palmlike tree
Also known as dwarf fan palm, the European fan palm often has a spreading and shrublike habit.

PROFILE

- Up to 40 ft (12 m)
- Up to 30 ft (10 m)
- Evergreen
- Whorled compound
- Canary Islands

leaflet up to 20 in (50 cm) long

dark green, pointed leaflet

Comblike leaves
The leaves are thick, leathery, comblike fronds. Individual leaflets are arranged in whorls around the stem.

FEATURES

Fruit Date-like, borne in bunches, green ripening to purple and then orange

Phoenix canariensis (Arecaceae)

CANARY ISLAND DATE PALM

This tree is native to the Canary Islands, and is one of the stateliest palms to thrive outside the tropics. It is surprisingly hardy and can withstand some frost. It has been widely cultivated as an ornamental species in warm, temperate regions throughout the world, particularly along seafronts and promenades.

The gray-brown bark has prominent scars left from where dead branches have been shed. The trunk on mature specimens may be up to 36 in (90 cm) wide and leaves are up to 15 ft (5 m) long. Golden yellow flower spikes, up to 6 ft (2 m) long, are borne in spring. These are followed by bunches of date-like fruit.

Stately palm
This palm has a round crown. Comblike fronds arch upward, away from the stout, straight trunk.

PROFILE

- Up to 25 ft (8 m)
- Up to 12 ft (4 m)
- Evergreen
- Whorled compound
- Argentina, Brazil, Uruguay

pinnate leaf, up to 10 ft (3 m) long

Distinctive leaves
The large, feathery leaves range in color from blue-gray to silver. The leaf stems have sharp, hooked spines along both edges.

FEATURES

Bark Gray-brown; patterned with persistent scars made by bases of previous leaves

Fruit Rounded, up to 1 in (2.5 cm) wide; borne in large clusters beneath the foliage

Butia capitata (Arecaceae)

JELLY PALM

Also known as pindo palm, this small palm is native to Argentina, Brazil, and Uruguay. However, because of its beauty and ability to withstand subzero temperatures, the jelly palm has been widely cultivated in the warm temperate regions of North America, Europe, and elsewhere. Its fruit is used to make jams, jellies (hence the name), and wine.

The jelly palm has a thick, sturdy trunk. Its small, fragrant flowers are yellow, tinged red to purple, and are carried in sets of threes—two male flowers and one female. These are followed by yellow or orange edible fruit known as pindo dates. They have a flavor like that of apricots, pineapples, and bananas mixed together.

Attractive crown
The jelly palm leaves arch upward and away from the stem, then curve at their tips back toward the trunk.

PROFILE

- Up to 30 ft (10 m)
- Up to 15 ft (5 m)
- Evergreen
- Whorled simple
- New Zealand

flower panicle bears individual flowers, ¼ in (5–6 mm) wide

Leaves and flowers
The branches bear erect leaves, which have a fibrous texture when torn. The long, pendulous flowers appear in spring and early summer.

FEATURES

Bark Gray, corky, spongy, shallowly fissured

Leaves Gray-green; up to 36 in (90 cm) long, 3 in (7 cm) wide

Cordyline australis (Asparagaceae)

CABBAGE PALM

This slow-growing, palmlike tree is not a true palm but more closely related to the agaves. The cabbage palm tree is native to New Zealand, where the indigenous Māoris used to eat the tender tips of its shoots, hence its common name "cabbage." Introduced into Europe in 1823, it has been widely cultivated in parks and gardens close to coastal regions.

It is usually a single-trunked tree with swordlike leaves that are striated with many parallel veins, and sometimes display a kink toward their tip. The palm bears masses of fragrant, white flowers in large panicles, 24–39 in (60–100 cm) long. The fruit is a small, pale lilac- to green-colored berry.

Ascending branches
The cabbage palm bears stout, ascending branches, which are crowned by a mass of swordlike leaves.

PROFILE

fruit up to 12 in (30 cm) long

Large fruit
The fruit is a hard, oval, green sphere, weighing up to 3 lb (1.5 kg). It appears amidst large, slightly arching, pinnate leaves.

- ↕ 70–100 ft (20–30 m)
- ↔ Up to 30 ft (10 m)
- 🌿 Evergreen
- ⬮ Compound
- ⊙ Pacific and Indian Ocean islands, N.W. South America, Australia, India

FEATURES

Bark Gray, marked with narrow leaf scars

seed 1½–3 in (4–8 cm) long

Seeds Brown, fibrous shell; containing edible white flesh, coconut milk, and embryo

Cocos nucifera (Arecaceae)

COCONUT

The origin of this popular palm remains a mystery. Some believe it is native to the Pacific and Indian Ocean islands; others suggest it was introduced from northwest South America. Fossils from around 40 million years ago place it in Australia and India. However, it is now widely spread across the tropics.

The trunk is long, slender, often curving. At the base, where new roots emerge, it may take on the shape of a hockey-stick hook. The leaves are up to 20 ft (6 m) long and consist of many lax, narrowly lanceolate leaflets, up to 20 in (50 cm) long. Creamy male and female flowers are carried in a 3 ft- (1 m-) long inflorescence with multiple branches.

"The tree of life"
The coconut tree is an important source of both food and medicine—hence its popular name.

PROFILE

- Up to 50 ft (15 m)
- Up to 30 ft (10 m)
- Evergreen
- Whorled simple
- Canary Islands, Cape Verde Islands, Madeira, W. Morocco

Dense foliage
Each branch of this short-trunked tree is topped by a rosette of thick, strap-like leaves, each ending in a narrow point.

blue-green leaf, up to 20 in (50 cm) long, 2 in (5 cm) wide

FEATURES

Bark Light brown, fades to gray in maturity; horizontally marked with old leaf scars

Dracaena draco (Asparagaceae)

DRAGON TREE

This widely spreading tree is indigenous to the Canary Islands, Cape Verde, Madeira, and western Morocco, and has also been planted in the Azores and around the Mediterranean region. The tree is also known as dragon's blood tree, because of its deep red sap, which was once used as a resin to stain wood.

The tree branches where it flowers. Two side shoots are borne at the base of each flower. After flowering, each shoot starts to grow and develops into a branch. The flowers are small, fragrant, greenish white, and borne in panicles (elongated flower clusters). These are followed by fleshy, bright orange berries, up to ½ in (1 cm) wide, each containing two seeds.

Wide-branched habit
The short trunk of the dragon tree divides into sturdy branches to form a dense umbrella-shaped canopy.

EUDICOTYLEDONS

Sometimes called eudicots, this is the biggest group of flowering plants and trees, encompassing an astonishing variety of woody seed trees. They are found throughout the world, and probably evolved more than 125 million years ago.

The term eudicotyledon comes from the embryonic seed leaves (called cotyledons) found in the seeds before they germinate. All eudicots have two or more cotyledons. Apart from this feature, these trees vary enormously. Some are evergreen, others deciduous; their leaves are simple or pinnate. They are found virtually worldwide: from temperate woodland to tropical rainforest, in arid deserts and at the edges of polar tundra.

Some of the most species-rich genera of trees are eudicots: oaks (*Quercus*) produce drab flowers in pendulous catkins for wind pollination; gum trees (*Eucalyptus*) use pollinating insects, birds or mammals—and have a correspondingly generous provision of nectar. Some of the most complex and highly evolved plant–insect interactions occur among eudicots: in the tropics, hundreds of species of fig trees (*Ficus*) use tiny wasps that complete their life cycles inside the fruit.

CLASS	ANGIOSPERMAE
SUBCLASS	EUDICOTYLEDONEAE
ORDERS	38
FAMILIES	307
SPECIES	182,227 (c.50,000 TREES)

MASTERS OF POLLINATION

Genetic similarities indicate that all eudicots evolved from a single common ancestor. The structural features that unite eudicots seem arcane but could have important biological significance. For instance, the pollen grains of all eudicots have three or more pores in their walls—unlike the grains of all other seed plants, which have one.

Female parts of most flowers have special receptive swellings—called stigmas—for catching pollen. Stigmas stimulate pollen tubes to grow through pollen pores, which guide male cells to the egg for fertilization. Stigmas of eudicots enjoy more effective pollination because the additional pores in the pollen make successful pollen tube growth more likely. Since these trees evolved, a combination of intimate flower–insect partnership and efficient pollen growth has led to an explosive diversification of eudicots. Today five in every six tree species are eudicots.

Gean (wild cherry)
Bee-pollinated flowers of this tree have both male and female parts; each forms a single seed encased in a stone.

Bright flowers
Rata has crimson bottlebrush-type flowers, which emerge from white-bloomed buds. Each flower bears long red stamens.

red stamen

young, elliptic leaf

PROFILE

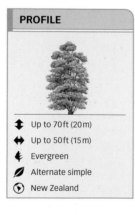

- ↕ Up to 70 ft (20 m)
- ↔ Up to 50 ft (15 m)
- Evergreen
- Alternate simple
- New Zealand

Metrosideros robusta (Myrtaceae)

RATA

A small to medium-sized, broadly columnar, evergreen tree, Rata is native to New Zealand, where it grows in close proximity to the sea. As well as being tolerant of sea air, it adapts well to wind and drought. This makes it an ideal ornamental tree for mild maritime districts. It was introduced into cultivation in Europe in 1840. Today, some of the finest European specimens of this tree grow on the island of Tresco, part of the Isles of Scilly, situated off the southwest coast of England.

Initially broadly columnar, this tree may become widely spreading in maturity. It has smooth, gray-brown bark. The leaves are initially narrow but become elliptic to almost rounded in maturity. They are dark green on the upper surface with a light green midrib. Thick and leathery in texture, leaves are covered with fine silver-gray hairs on the underside. The flowers appear in terminal clusters in summer, clearly visible in the dark, evergreen foliage.

dark green leaf up to 4 in (10 cm) long

light green leaf midrib

Christmas tree foliage
The thick, leathery leaves of this tree are narrowly oblong at first, becoming rounded in maturity. They have dense white hairs on the underside.

FEATURES

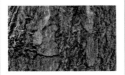

Bark Pale brown-gray, smooth; fissures into irregular plates in maturity

Flowers Bright crimson, bottlebrush-type in cultivation with long thin red stamens

Metrosideros excelsa (Myrtaceae)

NEW ZEALAND CHRISTMAS TREE

This dome-like, broadly spreading tree is native to New Zealand, where it flowers from November to January with a peak in mid-December—hence its common name. It has a natural range along the coastal fringes of New Zealand's North Island and the shores of the lakes in the Rotorua region. In the wild, it normally grows as a multi-stemmed tree, where its stems and branches may be covered in fibrous, aerial roots. This tree was introduced into western cultivation in 1840, but is relatively tender and only thrives in mild, coastal localities in Europe.

The tree flowers over two weeks and carries clusters of hermaphroditic flowers, which emerge from white-bloomed flower buds. In the wild, the flower color varies from bright crimson-red to pale pink, while in cultivation, the flowers are usually bright crimson. There is also a yellow-flowering cultivar called 'Aurea.'

Springtime growth
Arranged alternately along the shoot, young leaves mature by early summer. The flowers are arranged alternately.

young, rich green, ovate leaf

leaf bud

male flowers clustered together

lime-green young leaf

Up to 30 ft (10 m)

Up to 22 ft (7 m)

Deciduous

Alternate simple

China

FEATURES

Bark Grayish with deep, knobbly, crisscrossed fissures

long drawn-out leaf tip

serrated margin

Leaves Ovate, green, up to 6 in (16 cm) long

Eucommia ulmoides (Eucommiaceae)

EUCOMMIA

This attractive, unique species is the only hardy tree known to produce rubber. It is believed to be closely related to the elm—genus *Ulmus*—hence its species name *ulmoides*. The eucommia is a fast-growing, broadly spreading, deciduous tree and the only member of the Eucommiaceae family. It is thought to have grown wild in China, but no trees have been found outside of cultivation in recent years. It has been cultivated for centuries in China for its bark, which has medicinal properties, and was introduced into western cultivation in 1896. It is also grown as an ornamental or shade tree in temperate regions.

It has rich green leaves that are alternately arranged along the shoots. The leaves, if gently torn in half, will remain held together by thin, silklike strands of latex. The tree produces small, green-yellow, inconspicuous flowers in spring, and these are followed by oval, winged seeds ³⁄₁₆–1 ¼ in (2–3 cm) long and ¾ in (2 cm) wide—similar to those produced by elm trees.

*pale green
underside of leaf*

*long,
orange-red style*

*flower up
to 2 in
(5 cm) long*

*matte, blue-green
upper surface*

Tubular flowers
Red-orange tubular flowers are carried in clusters along the stems. They open into four narrow lobes, which peel back to reveal an orange-red style.

PROFILE

- Up to 28 ft (9 m)
- Up to 12 ft (4 m)
- Evergreen
- Alternate simple
- South America

Embothrium coccineum (Proteaceae)

CHILEAN FIRE BUSH

Originally from South America, this small, broadly columnar tree is fast becoming a popular species in gardens across the temperate world. It is hardy, and is often found growing in exposed locations in the wild, from the Pacific coast to high in the Andes mountains, both in Chile and Argentina. It was discovered and named from a specimen collected during Captain James Cook's second exploration voyage to this region in 1771 and introduced into cultivation by English plant collector William Lobb in 1846.

The bark is purple-brown, smooth when young, but flaking from the time the tree turns 30 years old. The leaves are elliptic, sometimes oblong to lanceolate, up to 6 in (15 cm) long and 1 in (2.5 cm) wide. The fruit is a woody capsule, up to 1 in (2.5 cm) long, with winged seeds.

Burning embers
The flowers of Chilean fire bush resemble burning embers from a distance, hence the name.

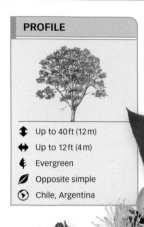

↕ Up to 40ft (12m)
↔ Up to 12ft (4m)
🌿 Evergreen
🗡 Opposite simple
⊙ Chile, Argentina

small, rounded flower bud

red-brown pedicel (flower stalk), up to ½in (1.5cm) long

oval to broadly elliptic leaf, up to 1in (2.5cm) long

Chilean myrtle flowers
The cup-shaped, white flowers are small and fragrant, held singly on stalks. They have four petals and numerous stamens.

FEATURES

Bark Bright cinnamon-orange, with soft, feltlike texture; peels in patches to reveal cream-colored surface beneath

Luma apiculata (Myrtaceae)

CHILEAN MYRTLE

Formerly known by the botanical name *Myrtus luma*, the Chilean myrtle is one of the most interesting trees for bark color. It has been widely planted in parks and gardens since its introduction in 1844 from the temperate forests of Chile and Argentina. It easily establishes itself in new locations, especially in western Europe, where the North Atlantic drift provides the warm, wet weather it enjoys. In southern Ireland, for example, the tree has become naturalized, reproducing itself with ease.

The tree has a broadly columnar habit. Curiously, its bark feels cold even when in full sunlight. The foliage is dark dull green, and the leaves emit a pleasant aroma when crushed. It bears flowers in profusion in late summer and fall. The rounded, fleshy fruit is up to ½in (1cm) long, and turns purple-black when ripe.

♤ **White Chilean myrtle (*Luma chequen*)** Besides being more shrublike in habit than its larger cousin, the white Chilean myrtle also has gray rather than cinnamon-orange bark.

PROFILE

⬍ Up to 80 ft (25 m)

⬌ Up to 50 ft (15 m)

🗲 Deciduous

🍃 Alternate simple

◉ E. North America

young leaf

flowers borne in drooping, axillary racemes

yellow anthers

small, greenish yellow male flower

nconspicuous bloom
lowers of the sassafras do not
ave petals. Male and female
owers are borne on separate
rees in late spring.

FEATURES

*leaf up to 6 in
(15 cm) long, 4 in
(10 cm) wide*

Leaves Shape varies from
unlobed to three-lobed

Sassafras albidum (Lauraceae)

SASSAFRAS

The natural habitat of this tree extends from Canada
to Florida, and then westward into Kansas and Texas. Its
leaves and bark are strongly aromatic and have been
used in the past by both Native Americans and British
colonists for their medicinal
properties. The bark of the root
has been used to make a drink
similar to beer.

Sassafras can produce varying
leaf shapes on the same tree;
from lobed and fig-like, to
obovate, without lobes. Its
leaves are deep green on the
upper surface and blue-green
on the underside. In fall, they
turn orange-yellow before being
shed. Female trees produce a
dark blue egg-shaped berry, up
to ½ in (1 cm) wide.

Wide form
Sassafras is a broadly
columnar, medium-sized,
suckering tree, with flexible
zigzag branching.

leaf up to 5 in (12 cm) long

glossy, bright green upper surface of leaf

Foliage and flowers
The leathery leaves are oblong to oblanceolate, alternately borne along shoots. Small flowers emerge in clusters of 6 to 10 in spring.

yellow-green flower, ¼ in (5 mm) wide

PROFILE

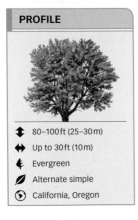

⬍ 80–100 ft (25–30 m)
⬌ Up to 30 ft (10 m)
🗲 Evergreen
🖊 Alternate simple
◉ California, Oregon

Umbellularia californica (Lauraceae)

CALIFORNIA LAUREL

Also known as the California bay, California olive, and Oregon myrtle, this is a vigorous, strongly aromatic species. It was introduced into western cultivation by Scottish plant collector David Douglas in 1829. This tree has a dense, leafy habit. The foliage, when crushed, emits a strong odor, which can induce nausea, headaches, and, in extreme cases, unconsciousness. It may also cause skin allergies. Its highly prized timber, known as pepperwood, has a pale brown color and can be polished to a fine finish.

The bark is dark gray and smooth when young, cracking into small, roughly square plates in maturity. The leaves, which resemble bay leaves, are borne on sage-green shoots. Its leaves are glossy bright green on the upper surface and yellow-green on the underside. Borne in umbels, inconspicuous flowers are held on each stem in the leaf axils. They are followed by small, pear-shaped berries, up to 1 in (2.5 cm) wide, ripening from green to dark purple.

PROFILE

↕ Up to 30 ft (10 m)

↔ Up to 25 ft (8 m)

🕭 Evergreen

▨ Opposite simple

◉ Mediterranean region, N. Africa, Middle East, China

gray-green leaf upper surface

small, white, cross-shaped flower, up to ¼ in (5 mm) wide

leaf up to 4 in (10 cm) long, 1¼ in (3 cm) wide

silver sage-green leaf underside

Leaves and flower clusters
The olive has tough, leathery, and narrowly obovate leaves and bears small racemes (stalks) of fragrant white flowers in summer

FEATURES

Bark Gray; fissured in youth

fruit up to 1¼ in (3 cm) long

Fruit Oval, green, edible; ripening to glossy purple-black

Olea europaea (Oleaceae)

OLIVE

One of the earliest trees to be cultivated by man, the olive tree has been highly prized for its fruit and oil for at least 5,000 years. Several specimens in Italy and Greece are more than 1,000 years old. Its natural range extends from the Iberian Peninsula, around the Mediterranean Sea, and eastward into China. It is cultivated commercially in warm regions elsewhere, including South Africa, Australia, and California. Although the tree survives in cooler climates, it seldom flowers or produces fruit.

The olive is small and evergreen. It has a short trunk, which becomes deformed, thick, and hollow with age. Its broadly spreading crown provides welcome shade in hot climates. Clusters of small flowers bloom in summer and are followed by a berrylike fruit, containing a single seed, or stone, from which olive oil is extracted.

♀ **Chinese fringe tree (*Chionanthus retusus*)** This related species is a deciduous, almost shrublike tree with showy clusters of snow-like flowers.

Shining privet flowers
Small, creamy white flowers are borne in large clusters, or panicles, which stand proud above the surrounding foliage.

flowers borne in large panicle, 8 in (20 cm) long

ovate leaf, up to 4 in (10 cm) long, 2 in (5 cm) wide

glossy green upper surface on leaf

pale green, matte underside of leaf

PROFILE

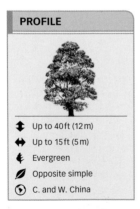

↕ Up to 40 ft (12 m)
↔ Up to 15 ft (5 m)
🌿 Evergreen
🍃 Opposite simple
⏱ C. and W. China

Ligustrum lucidum (Oleaceae)

SHINING PRIVET

Also known as Chinese privet, this tree is indigenous to mountains in central and western China, where it grows in hillside woodland above river valleys. It is believed to have been introduced into the West by the botanist Joseph Banks in 1794. It is now commonly planted as a street tree in urban areas because it is able to tolerate atmospheric pollution, compacted dry soils, and regular pruning.

The gray bark is smooth even in maturity, although the trunk may become fluted in old age. The ovate leaves taper to a long, fine point, and are bronze when first produced. The flowers emerge in late summer and early fall, followed by clusters of small, blue-black, rounded berries in late fall.

Lush evergreen tree
The medium-sized shining privet tree produces a dense canopy of lush, lustrous leaves.

PROFILE

↕ Up to 30 ft (10 m)

↔ Up to 25 ft (8 m)

🍂 Evergreen

🌿 Opposite simple

⊙ S. Europe, Canary Islands

Jasmine flowers
These fragrant flowers are small, cream or yellowish white, and are produced in clusters in the leaf axils in spring.

flower has yellow anthers

leaf up to 2 in (5 cm) long

bluntly toothed leaf margin

FEATURES

Bark Smooth, pewter-gray throughout life, relatively thin

Fruit Blue-black, up to ½ in (1 cm) wide

Phillyrea latifolia (Oleaceae)

JASMINE BOX

A member of the olive family, this small, broadly spreading tree is often mistaken for the holm oak (p.167), which originates from the same region. Jasmine box has been in cultivation for at least 400 years, and is widely planted to provide shelter in exposed conditions, especially along the coast. It produces masses of dark green, luxuriant foliage, in some cases causing the branches to bow down under the weight of the leaves, and is often clipped to form a dense hedge, or for topiary.

The thick and leathery leaves are elliptic to ovate, borne opposite each other. They have a glossy dark green upper surface, a pale green underside, and bluntly toothed margins. The flowers bloom in spring and are sometimes followed by berrylike fruit.

◊ **Mayten**
This tree (*Maytenus boaria*) has a graceful habit, slender branches, and narrow, elliptic, finely toothed leaves.

reddish purple flower
up to ¼ in (7 mm) wide

leaf up to 2 in
(5 cm) long,
1 in (2.5 cm) wide

pale green to
white underside
of leaf

white midrib on
upper surface of leaf

Glossy green leaves
The are shiny and
slightly waxy with wavy
edges. Their shape varies
between oblong and elliptic.

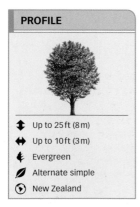

PROFILE

↕ Up to 25 ft (8 m)

↔ Up to 10 ft (3 m)

⚘ Evergreen

🖊 Alternate simple

⊙ New Zealand

Pittosporum tenuifolium (Pittosporaceae)

KOHUHU

This small, columnar tree is indigenous to both the
North and South islands of New Zealand. It has been widely
cultivated in Europe and elsewhere for the last 160 years,
even though its foliage is prone to scorch and browning
in cold winters. Kohuhu is a Māori name, which roughly
translates as "the black one" and refers to the color of
the bark, twigs, and inner wood. The ornamentally placed
leaves make this tree's foliage particularly attractive to
flower arrangers and florists.

The bright green leaves are alternately placed on thin,
dark purple to almost black twigs. The conical leaf bud is
pale yellow-green, tipped with crimson-purple, and up to
1⁄16 in (2 mm) long. The small, dark flowers are cup-shaped
and bear five petals with out-curved tips. They appear in
the leaf axils from late spring to early summer, and are
particularly fragrant at night.

◊ **Karo (***Pittosporum crassifolium***)** This tree bears large,
leathery, oval-shaped leaves, and purple flowers that are
larger and more showy than that of the kohuhu.

PROFILE

- Up to 100ft (30m)
- Up to 70ft (20m)
- Deciduous
- Alternate simple
- W. Europe

Dutch elm foliage
The leaves are rough and unequal at the base, with hairs on the underside. A pink stalk attaches the leaves to the shoot.

fruit up to 1 in (2.5 cm) long

leaf up to 5 in (12.5 cm) long, 3 in (7.5 cm) wide

irregularly serrated leaf margin

oval to ovate-acuminate, green-gray leaf

FEATURES

Bark Gray-brown; smooth at first, becomes vertically cracked in maturity

Ulmus x hollandica (Ulmaceae)

DUTCH ELM

This broadly columnar tree, also known as the European hybrid elm, is a natural hybrid between the smooth-leaved elm (*Ulmus minor*) and the wych elm (p.136). It is found throughout western Europe, including Britain, and is widespread in France, Belgium, Germany, and the Netherlands.

The spreading canopy of the Dutch elm is open and thin, with relatively few branches ascending steeply away from the trunk. The base of the tree's trunk may be fluted, and its twigs may have corky ridges. Dark red flowers are produced in large clusters on stout shoots in early spring. The fruit is an oval, winged seed; it is pale green turning to gold and, finally, brown in early summer.

⌂ **Smooth-leaved elm**
Related to the Dutch elm, this tree (*Ulmus minor*) has smaller leaves with a smooth upper surface.

Elm leaves
The leaves of the European white elm are obovate and distinctly oblique at the base, with double-toothed margins.

leaf 4½in (11cm) long, 2½in (6cm) wide

double-toothed leaf margin

PROFILE

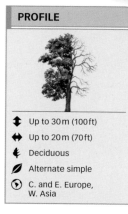

↕ Up to 30m (100ft)

↔ Up to 20m (70ft)

🍂 Deciduous

🌿 Alternate simple

◉ C. and E. Europe, W. Asia

FEATURES

Bark Remains smooth, ridged, even in old age

Fruit Oval-shaped, paper-like samara; contain single seed, green ripening to brown

Ulmus laevis (Ulmaceae)

EUROPEAN WHITE ELM

This large, widely spreading tree is native to central and eastern Europe and western Asia. It has been widely cultivated for many years, particularly in Russia, where it is planted alongside roads and railway lines to protect them from snowdrifts. The other name for this majestic tree is fluttering elm, which refers to the fact that the long-stalked flowers and fruit flutter in the slightest breeze.

Whereas most other elms develop a rough, fissured bark, the bark of this tree remains smooth, even in maturity. This gives this tree its species name *laevis*, which translates as "smooth." The leaves are mid-green on the upper surface, with soft, gray down on the underside. There may be up to 17 veins on the long side of the leaf and only 14 on the short side. Small, purple flowers emerge in clusters in early spring. In summer, fruit are borne in bunches on long, pendulous stalks.

◌ **Siberian elm (*Ulmus pumila*)** This tree, up to 30ft (10m) tall, is distinguished by its smaller size, multi-stemmed often shrubby habit, and small, ovate to lanceolate leaves, up to 2in (5cm) long.

Coarse leaves
The ovate leaves are pointed at the tip, unequal at the base, and rough to the touch. The fruit are small, and oval-shaped structures containing the seed.

leaf up to 4 in (10 cm) long, 3 in (7 cm) wide

pale, membranous wing

coarsely toothed leaf margin

winged fruit up to 1 in (2.5 cm) long

PROFILE

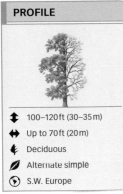

- 100–120 ft (30–35 m)
- Up to 70 ft (20 m)
- Deciduous
- Alternate simple
- S.W. Europe

Ulmus procera (Ulmaceae)

ENGLISH ELM

Native to southwestern Europe, this large, broadly columnar tree was once an important part of the landscape in England, and featured widely in the paintings of English landscape artists such as John Constable. Dutch elm disease led to its demise from the 1960s to 1980s. It is now believed that the English elm is not English at all, and was introduced into Britain from northwest Spain in the early Neolithic period.

The English elm has a gray-brown bark, which becomes regularly fissured into rough rectangular plates from an early age. The leaves are dark green on the upper surface and pale green on the underside, each with 10 to 12 pairs of veins. They turn golden yellow in fall before being shed. Small clusters of red flowers appear on bare shoots in early spring, followed by flattened, oval-shaped fruit (samara) containing paper-thin, winged seeds in summer.

Japanese elm (*Ulmus japonica*) With a more rounded, broadly spreading habit and larger leaves, this tree easily stands out from its cousin, the English elm.

PROFILE

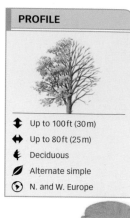

- Up to 100 ft (30 m)
- Up to 80 ft (25 m)
- Deciduous
- Alternate simple
- N. and W. Europe

Papery fruit
The wych elm fruit is a papery, hairy disc known as a samara, borne in clusters, which carries seeds.

seed carried in center of disc

disc up to ¼ in (5 mm) wide

leaf up to 8 in (20 cm) long, 4 in (10 cm) wide

FEATURES

Bark Gray-brown, smooth; becomes lined with vertical fissures on maturity

double-toothed leaf margin

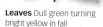

Leaves Dull green turning bright yellow in fall

Ulmus glabra (Ulmaceae)

WYCH ELM

This medium to large, broadly spreading, deciduous tree is found throughout western and northern Europe, from Russia to Portugal and north to Scandinavia. It survives well in exposed coastal areas and on low mountain slopes, especially in areas with high rainfall. It has strong and dense timber, which is resistant to decay by water. For centuries, hollowed out elm branches were used as water conduits, while sawed planks from the trunk were used to make coffin boards.

The wych elm has a short trunk that rapidly divides into large, ascending branches. Bright, red-purple flowers emerge in early spring before the leaves unfurl. The oval to obovate leaves are dull green and rough on the upper surface and pale green on the underside, with gray hairs. They are coarsely and doubly toothed at the margins, and have unequal sides at the base.

♧ **Chinese elm (*Ulmus parvifolia*)** With its small, leathery, and glossy green leaves, the Chinese elm is relatively easy to distinguish from the wych elm.

sharply toothed
leaf margin

leaf up to 6 in (15 cm)
long, 3 in (7 cm) wide

three-veined
base

Narrowly oval leaf
The large, lanceolate to ovate leaves
of the southern nettle tree taper to
pointed tips and turn pale yellow in
fall before being shed.

PROFILE

↕ Up to 80 ft (25 m)

↔ Up to 70 ft (20 m)

🌿 Deciduous

▱ Alternate simple

⊙ S. Europe, Middle East,
N. Africa

FEATURES

Bark Smooth and gray;
develops corky warts and
ridges with age

flower
without
petals

Flowers Small and green;
borne either singly or in small
clusters in spring

Celtis australis (Ulmaceae)

SOUTHERN NETTLE TREE

Also called European nettle tree or Mediterranean
hackberry, this broadly spreading tree has been cultivated
in the British Isles since the 16th century. It is often planted
for ornamental purposes and is resistant to air pollution. Its
sweet and edible fruit can be eaten raw or cooked. Both
leaves and fruit are believed to have some medicinal
properties and are used to treat symptoms such as
abdominal pain and colic.

The large leaves are rough and have a dark gray-green
upper surface and a paler underside with a covering of
bristly hairs. The flowers are hermaphroditic, with both
male and female organs within each flower. Ovoid,
purple-brown to blackish berrylike fruit are borne in short
clusters and eaten by birds and other wildlife.

◌ **Japanese hackberry (*Celtis sinensis*)** Reaching up to only
50 ft (15 m) in height, this tree is smaller than the southern nettle
tree with smoother, smaller, glossy leaves.

PROFILE

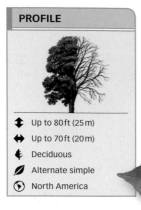

- Up to 80 ft (25 m)
- Up to 70 ft (20 m)
- Deciduous
- Alternate simple
- North America

leaf margin sharp-toothed near tip

rounded fruit, up to ½ in (1 cm) wide

leaf up to 5 in (12 cm) long and 2 in (5 cm) wide

Foliage and fruit
Hackberry leaves are oval to ovate and pointed at the tip. The fruit is a small berry, turning from green to red, then ripening to purple-black.

FEATURES

small, green flower

Flowers Without petals; borne in leaf axils

Celtis occidentalis (Ulmaceae)

HACKBERRY

Also known as the nettle tree, the leaves of the hackberry bear a distinct resemblance to the leaves of the common stinging nettle (*Urtica dioica*). This medium-sized, broadly columnar, deciduous tree is native to most of North America. It has been widely cultivated in Europe and elsewhere since the 17th century.

The bark is light gray, smooth when young, developing wart-like blemishes in maturity. The leaves are oval to ovate and roughly serrated around the margin, except for the lower third of the leaf. They are smooth, glossy, and rich green on the upper surface, and lighter green with slightly hairy veining on the underside. Male and female flowers are held separately on the same tree in spring. They are followed by fruit that are borne singly on thin, green stalks up to 1 in (2.5 cm) long.

◊ **Mississippi hackberry (*Celtis laevigata*)** This tree is distinguished from the hackberry by its narrower, lanceolate leaves that are rarely serrated around the margins.

Vibrant green foliage
Elliptic to ovate leaves are borne on hairy shoots. The leaves have distinct parallel veining, a bright green upper surface, and a pale green underside.

coarsely toothed
leaf margin

leaf has
rough texture

leaf up to 4 in
(10 cm) long

PROFILE

- ↕ Up to 100 ft (30 m)
- ↔ Up to 50 ft (15 m)
- Deciduous
- Alternate simple
- Caucasus region, N. Iran

FEATURES

Bark Smooth, silver-gray when young; flakes in old age

Zelkova carpinifolia (Ulmaceae)

CAUCASIAN ELM

Related to the genus *Ulmus*, this slow-growing, long-lived forest tree grows up to 100 ft (30 m) tall. The Caucasian elm is widely planted as an ornamental tree in parks, gardens, and arboreta in western Europe. It bears some resemblance to the common hornbeam (p.184), with a distinctive rounded shape, a short trunk, and a large mop head of dense, ascending branches. The overall impression is that of a tree that was once pollarded (in which the top of the tree is cut off to encourage branching) but has long since been left unchecked.

The smooth bark is like that of the common beech (p.152) but flakes in maturity. The trunk has pronounced fluting with a buttressed base. Leaves are vibrant green and rough to the touch. Small green flowers are borne in the leaf axils in spring. The fruit is a small, rounded, pealike capsule with pronounced ridging.

◊ **Chinese zelkova (*Zelkova sinica*)** This is a small tree, up to 30 ft (10 m) tall, with slender branches and small, ovate to lanceolate leaves, 1–2½ in (2.5–6 cm) long.

Keyaki leaves
The leaves have 6 to 13 teeth on each side. They have a dark green upper surface and a paler underside with hairs.

sharply toothed leaf margin

leaf up to 5 in (12 cm) long, 2 in (5 cm) wide

leaf ends in slender tip

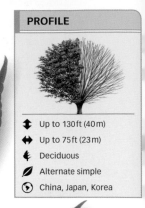
FEATURES

Bark Pale gray; peels or flakes in maturity to reveal light fawn-gray patches beneath

Zelkova serrata (Ulmaceae)

KEYAKI

Native to China, Japan, and Korea, the broadly spreading keyaki grows in low-lying river valleys, where the soil is rich and deep. It was introduced into western cultivation in 1861, and since then it has become a popular tree for planting in large gardens and parks. In Japan, it grows as a forest tree at elevations of up to 4,000 ft (1,200 m) above sea level. Many old Japanese temples, especially in the Kyoto region, are built of its strong and durable timber.

The smooth bark is rather like that of the common beech (p.152). The ovate leaves are slightly rough on the upper surface. They turn orange-red in fall before being shed. Small green male and female flowers appear in spring, followed by rounded fruit, up to ¼ in (5 mm) wide.

Rounded crown
An elm-like tree, the keyaki has light, ascending lower branching and a rounded crown.

leaflet up to 5in
(13cm) long

PROFILE

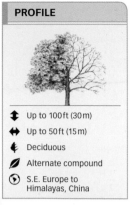

↕ Up to 100ft (30m)

↔ Up to 50ft (15m)

🍂 Deciduous

🌿 Alternate compound

🔄 S.E. Europe to
Himalayas, China

fruit up to 2in
(5cm) wide

mature leaflet, deep
green with dull sheen

Well-known nut
The fruit that encases the English
walnut is a round, green husk
carried on a short, thick stalk.

FEATURES

Bark Light gray; develops
narrow, vertical ridges with age

smooth,
hard
husk

Seeds Brown walnut;
containing edible kernels

Juglans regia (Juglandaceae)

ENGLISH WALNUT

This relatively slow-growing, medium-sized to large
deciduous tree is broadly spreading. It is indigenous to
Greece through the Himalayan region to China. It has
been widely cultivated elsewhere for its nuts and highly
prized decorative timber since the Roman Empire. It was
the Romans who introduced it to the British Isles. It
was in cultivation in the US by the 18th century.

The English walnut has a
rounded canopy of long,
spreading branches. Its aromatic
leaves are pink-bronze when
young. They consist of up to nine
leaflets, growing from a main leaf
stalk. Male and female flowers
are borne on the same tree, both
long, green catkins that appear
in spring. Its fruit is a round,
green husk, which may stain
when handled.

◇ **Butternut**
This tree (*Juglans cinerea*)
has purple-gray bark and
larger leaves than the
English walnut.

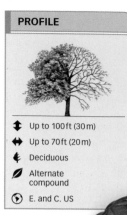

↕ Up to 100 ft (30 m)

↔ Up to 70 ft (20 m)

🍂 Deciduous

🌿 Alternate compound

⊙ E. and C. US

Glossy leaflets
The pinnate leaves consist of up to 17 leaflets, with toothed margins and pointed tips. The upper surface is glossy dark green, while the underside is hairy.

leaflet up to 4 in (10 cm) long

yellow-green male catkin, up to 4 in (10 cm) long

FEATURES

green husk

Fruit Rounded, smooth husk; encasing edible nut

Juglans nigra (Juglandaceae)

BLACK WALNUT

Native to the eastern and central US, this fast-growing tree has been widely cultivated in large gardens, parks, and arboreta in the temperate world. It has a pyramidal habit that spreads with age, and large leaves that turn bright yellow in fall before shedding. The tree is also prized for its timber, which is used for making cabinets.

The smooth, dark gray-brown bark develops fissures and ridges after about 30 years. Both male and female flowers are small and petal-less, appearing on the same tree in catkins in late spring and early summer. The gray-green female flowers are borne near the ends of shoots. The fruit is not as large as the English walnut (p.141), but has the same distinctive flavor.

Late to leaf
The broadly spreading black walnut's leaves appear late, and the tree remains bare till early summer.

leaflet up to 6 in
(15 cm) long

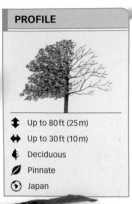
red stigma on
female flower

slightly toothed
margin

sticky brown
shoot covered
with hairs

Leaflets and flowers
The leaflets have a slightly hairy upper
surface and a furry underside. Female
catkins are similar to males, but can be
told apart by their distinct red stigmas.

FEATURES

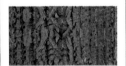

Bark Gray-brown; becomes
fissured and segmented into
irregular plates

sticky,
hairy husk

Fruit Green and rounded,
sometimes oval; containing
edible nut

Juglans ailantifolia (Juglandaceae)

JAPANESE WALNUT

As the name suggests, this medium-sized, deciduous
tree is native to Japan, but it has been planted elsewhere
for ornamental purposes. It has the largest leaves among
the walnuts and, when viewed from a distance, a denser
appearance. It is a fast-growing species, with long
spreading branches.

Each pinnate leaf may be up to 32 in (80 cm) long, and
consists of up to 17 leaflets. Male and female flowers—
both pendulous, green catkins—
are borne separately on the same
tree in late spring. Fruit are
carried in clusters of up to 20,
each a rounded, green husk
containing the nut. The sticky
husk is poisonous, and was
used by the Japanese to catch
fish. However, the walnut is
edible, and has been eaten in
Japan for many centuries.

Ornamental tree
The Japanese walnut tree
is conical when young, but
spreads with age. Its leaves
are glossy, dark green.

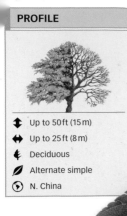
Fruit and leaves
Borne in short clusters, the white mulberry's fruit look like elongated raspberries. The leaves are variable in shape, usually ovate to rounded.

deeply toothed leaf margin

fruit up to 1 in (2.5 cm) long

leaf up to 8 in (20 cm) long, ending in long tip

FEATURES

Bark Brown-orange, becomes gnarled from early age

Fruit Turn from green to red

Morus alba (Moraceae)

WHITE MULBERRY

A widely spreading, medium-sized tree, the white mulberry is best known for being the food source of silkworms, which devour its leaves with gusto. Native to northern China, it is believed to have been introduced into Europe before 1596. Since then, it has become widespread across the temperate world.

The white mulberry's leaves are bright green and lustrous on the upper surface and paler on the underside. In fall, they turn rich butter-yellow before being shed. In some parts of the world, the leaves are used as fodder. The fruit can vary in color, from white to pink to purple-red. It is edible, though usually less sweet and produced in less profusion than those of its more popular relative, the black mulberry (p.145).

♤ **Paper mulberry**
This tree (*Broussonetia papyrifera*) can be told ap~ by its spherical female flowers and orange-red

PROFILE

↕ 25–30 ft (8–10 m)
↔ 25–30 ft (8–10 m)
🍂 Deciduous
🌿 Alternate simple
🌐 W. Asia

Edible berry
Black mulberry fruit appear at the end of twigs, and ripen from green through red to dark purple. The leaves are rough and hairy.

bright green, ovate leaf with coarsely serrated margin

leaf underside is lighter green and hairy

barely stalked fruit, borne in clusters

FEATURES

clustered male flower

Male flowers Small, green, cone-like structures

Morus nigra (Moraceae)

BLACK MULBERRY

Originally from western Asia, the black mulberry quickly spread through human intervention and cultivation across temperate Asia and Europe. It reached the British Isles in the early 16th century, and North America thereafter. This tree has been popular for its edible fruit. A long-lived tree, several examples are known to be over 500 years old. Its orange-brown bark becomes fissured and cracked from an early age.

Its leaves are 5 in (12 cm) long and 4 in (10 cm) wide. Male and female flowers are normally borne on separate trees in early summer. These are followed by raspberry-like fruit on female trees. The fruit is edible when ripe and has an astringently sweet taste.

Gnarled habit
A broadly spreading tree, the black mulberry has an old appearance from a relatively early age.

PROFILE

- Up to 30 ft (10 m)
- Up to 30 ft (10 m)
- Deciduous
- Alternate simple
- S.W. Asia

Distinctive leaves and fruit
Fig leaves have three to five deep lobes—the middle lobe being the largest. The fruit are borne near the shoot tips.

coarsely and bluntly toothed leaf margin

leaf up to 12 in (30 cm) long, 10 in (25 cm) wide

green, unripe fruit

FEATURES

fleshy receptacle contains several tiny seeds

Fruit Pear-shaped, up to 2 in (5 cm) long; swell and ripen to purple-brown in second year

Ficus carica (Moraceae)

COMMON FIG

One of the earliest cultivated trees, the small, shrubby, but broadly spreading common fig originates in southwest Asia. However, it is grown throughout warm regions across the world for its edible fruit. The fruit is also used in Greco-Roman, Islamic, Ayurvedic, and traditional Chinese medicine. In cooler regions, it is usually grown against a wall rather than as a freestanding tree.

The bark is mid-gray and smooth, with some patterning in old age. The trunk is short, sometimes twisted, while the branches are long and sinuous, starting from low down on the trunk and ending in stout and knobbly, upswept shoots. The easily recognizable leaves are large, leathery, and rough, with a shiny dark green upper surface and a paler, hairy underside. They have prominent veins and are heart-shaped at the base. Its uniquely shaped fruit are produced in summer near the tips of the shoots. Green in the first year, the fruit ripen over two years.

PROFILE

↕ Up to 50 ft (15 m)

↔ Up to 30 ft (10 m)

🍂 Deciduous

🍂 Alternate simple

⊙ C. and S. USA

Untooothed leaves
The leaves are ovate, pointed at both ends, and untoothed. They are glossy, rich green on the upper surface and pale green on the underside.

smooth leaf surface

leaf up to 4 in (10 cm) long, 2 in (5 cm) wide

leaf midrib

FEATURES

Bark Orange-brown; with scaly ridges and uneven furrows

fruit up to 4 in (10 cm) wide

Fruit Green ripening to yellow; with an even, stippled surface and occasional brown hairs

Maclura pomifera (Moraceae)

OSAGE ORANGE

Native to the central and southern United States, this broadly spreading, ornamental tree was introduced into western cultivation in 1818. In the wild in the US, it grows primarily in wet areas alongside rivers. Parts of the tree are used in food processing, pesticide production, and dye-making.

The tree has fleshy, yellow roots and a short trunk. Branches are often twisted and, when broken, ooze out a milky sap. They have sharp, green spines, normally at the base of leaf stalks. Small yellow-green male and female flowers, up to ½ in (1 cm) long, are borne in clusters on separate trees in early summer. The male flowers are borne in short racemes, while the females appear in clusters up to ½ in (1 cm) long. Female flowers ripen to become a large, spherical syncarp (multiple fruit), weighing up to 2¼ lb (1 kg). The tree is probably best known for this showy fruit, which looks similar to an orange. However, the fruit is inedible and, when fresh, full of sour milky juice.

- Up to 100 ft (30 m)
- Up to 50 ft (15 m)
- Deciduous
- Alternate compound
- S.E. United States

Pinnate leaves
Leaves of the bitternut hickory consist of up to four pairs of opposite leaflets and one terminal leaflet.

leaflet up to 15 cm (6 in) long, 7.5 cm (3 in) broad

leaflet has a serrated margin

green male catkin, up to 3 in (7.5 cm) long

FEATURES

Bark Gray and smooth; becomes thick, fissured, and ridged in maturity

Fruit Oval, thin, green shell; encasing bitter nut

Carya cordiformis (Juglandaceae)

BITTERNUT HICKORY

This large, broadly columnar, deciduous tree is native to the southeastern US, from Minnesota to Texas. It is planted elsewhere in parks, gardens, and arboreta for its fall leaf color and ornamental ridged, gray bark. As the name suggests, its fruit are hard nuts and not palatable. They do, however, have a high oil content, and were once crushed to produce lamp oil. The Native American word for hickory is "Pawcohiccora," which means nut oil.

A good way to distinguish this species from other hickories is its winter buds, which are long, slender, and bright yellow. Leaves are deep green on the upper surface and yellow-green on the underside. They turn golden yellow in fall. Both male and female flowers are carried on separate catkins on the same tree in late spring. Male catkins ripen from green to gold.

◇ **Pignut (*Carya glabra*)** Although its flowers and fruit are similar to the bitternut hickory, this tree is smaller, up to 70 ft (22 m) tall. Its orange-brown leaf bud is up to ½ in (0.8 cm) long.

PROFILE

- Up to 70 ft (20 m)
- Up to 40 ft (12 m)
- Deciduous
- Alternate compound
- E. United States

yellow-green leaflet

terminal leaflet

shallow-toothed leaf margin

Shellbark hickory leaf
The large compound leaf of this tree normally consists of seven leaflets, three opposite pairs and one terminal leaf.

leaflet up to 12 in (30 cm) long, 6 in (15 cm) wide

FEATURES

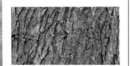

Bark Light gray, initially smooth; quickly develops shallow orange fissures

Fruit Yellow-green; borne in pairs or threes

Carya laciniosa (Juglandaceae)

SHELLBARK HICKORY

Also known as kingnut hickory, this tree was introduced into European cultivation in 1804. This slow-growing, medium-sized tree has a broadly columnar habit. It is now widely planted in ornamental collections for its large, attractive foliage and fall leaf color. The Latin name *laciniosa* means "with flaps," and refers to the way the shaggy bark curls away from the trunk.

The bark peels in thick vertical strips. Large, ovoid, bright leaf buds, up to 1 in (2.5 cm) long, open in spring to release the largest compound leaf of any hickory. The lanceolate to obovate-lanceolate leaves are up to 30 in (75 cm) long. The thick, tough leaflets are pale green with slightly hairy veins on the underside. They turn golden yellow in fall. The big pear-shaped or rounded fruit are up to 2 in (5 cm) long, with a shell covering a sweet, edible, large nut.

◊ **Pecan (*Carya illinoinensis*)** This tree is distinguished from the shellbark hickory by the number of leaflets on each pinnate leaf, usually 11 to 17, and its narrower, oval fruit.

terminal leaflet up
to 8 in (20 cm) long,
4 in (10 cm) wide

leaflet
tapers
towards tip

toothed
leaflet margin

Pinnate leaves
The leaves are more than 12 in (30 cm) long, with up to nine obovate leaflets. They turn gold in autumn.

FEATURES

Bark Silver-gray; initially smooth, develops vertical orange fissures with age

fine hairs
on leaf bud

Leaf buds Ovoid; fade from red-brown to pale fawn

Fruit Rounded to oval, yellow-green husk

Carya tomentosa (Juglandaceae)

BIG-BUD HICKORY

Also known as mockernut, this broadly columnar, North American hickory is native east of a line from Minnesota to Texas. In the wild, it is the most abundant and widespread of the hickories. It is long lived, sometimes reaching 500 years old. The species name comes from the Latin word *tomentum*, meaning "covered with dense, short hairs." This refers to the underside of the leaves and helps to identify the species. The tree is prized for its timber, which is strong and can withstand impact, making it the perfect wood for tool handles and sports equipment such as hockey sticks. When burnt, the wood emits a fragrant smoke, which is used to smoke meat.

As the name implies, the big-bud hickory has a large leaf bud, which is up to ¾ in (2 cm) long. The leaves have a slight sheen and are dark green on the upper surface, with a paler underside. They emit a fragrance when crushed. Both the shoot and the leaf stalk are covered in fine, fawn-colored hairs. The fruit has seams that resemble those of an old-fashioned football.

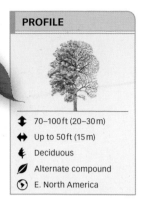

↕ 70–100 ft (20–30 m)

↔ Up to 50 ft (15 m)

🕯 Deciduous

🌿 Alternate compound

◐ E. North America

obovate leaflet, up to 4 in (10 cm) long

catkins clustered in groups of three

top two-thirds of leaflet serrated at margin

Long leaves
The pinnate leaves are made up of five to seven leaflets, each with a long tip.

FEATURES

Bark Shaggy gray; curls in narrow, vertical strips

Fruit Oval to rounded, up to 1¼ in (3 cm) long; enclosing edible nut

Carya ovata (Juglandaceae)

SHAGBARK HICKORY

This is the most valuable nut-producing tree in the US, with a natural range extending from Quebec, Canada, to Texas. It was one of the first North American trees to be introduced into Europe, arriving in 1629, and is widely planted for its fall tints.

As the name suggests, it has shaggy, gray bark, which curls away from the tree in thin strips up to 12 in (30 cm) long, staying attached to the tree at the center. This gives the trunk a rather disheveled appearance. The pinnate leaves are deep yellow-green, turning brilliant golden yellow in fall before being shed. Small, yellowish green male and female flowers are borne on pendulous catkins in late spring. The thin, green husk of the fruit encases a sweet-tasting nut.

Vigorous habit
The shagbark hickory has a dense, broadly columnar habit and is widely planted for its fall coloration.

Beech nut and leaves
A woody, brown husk covered with coarse bristles carries up to three triangular edible nuts.

- Up to 130 ft (40 m)
- Up to 70 ft (20 m)
- Deciduous
- Alternate simple
- Europe, British Isles to Russia

brown, woody husk

wavy-edged leaf, untoothed around margin

leaf up to 4 in (10 cm) long, 2 in (5 cm) wide

FEATURES

Bark Silver-gray; smooth throughout the tree's life

Fagus sylvatica (Fagaceae)

COMMON BEECH

Also known as European beech, this majestic, broadly spreading tree takes its name "beech" from the Anglo-Saxon *boc* and the Germanic word *Buche*, both of which mean "book." In northern Europe, early manuscripts were written on thin tablets of beech wood and bound in beech boards. Nowadays, it is widely used for hedging because it retains its dead brown leaves throughout winter, thereby providing additional shelter.

Leaves are ovate to obovate, dark green on the upper surface and slightly paler beneath. New leaves are yellow-green, covered in fine hairs, and are edible, with a nutty flavor. Older leaves are tough and bitter. Yellow-green flowers appear in spring. Both male and female flowers are borne in separate clusters on the same tree.

△ American beech
This tree (*Fagus grandifolia*) is distinguished from common beech by its longer, narrower leaves and its suckering habit.

PROFILE

↕ Up to 130 ft (40 m)

↔ Up to 70 ft (20 m)

🌿 Deciduous

▨ Alternate simple

◉ Switzerland, E. France

pod covered with bristles

smooth, wavy leaf margin

Foliage and fruit
The leaves range in color from purple-green to deep purple, almost black. The fruit is a small, oval pod covered with bristles.

leaf up to 3½ in (9 cm) long, 1½ in (4 cm) wide

FEATURES

Bark Silver-gray, smooth throughout the life of the tree

Fagus sylvatica Purpurea Group (Fagaceae)

PURPLE BEECH

This large, broadly spreading tree is neither a true species nor a garden creation, but rather a quirk of nature. Purple-leaved beeches were first spotted near Buchs, Switzerland, and in the Darney forest in eastern France around 1600. Shortly afterward, they started being cultivated as an ornamental oddity, first in those regions and then farther afield, soon becoming a favorite for planting in parks.

The purple beech is similar to common beech (p.152) in terms of its smooth, silver-gray bark and untoothed, wavy leaf margin. However, instead of being green, the leaves span various shades of purple—from vibrant pink-purple when young, to dark purple in maturity. They are also more oval and slightly smaller, with blunt tips.

⬙ **Weeping beech**
This tree (*Fagus sylvatica* 'Pendula') is distinguished from the purple beech by its rich green leaves.

PROFILE

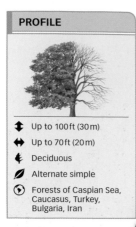

- Up to 100 ft (30 m)
- Up to 70 ft (20 m)
- Deciduous
- Alternate simple
- Forests of Caspian Sea, Caucasus, Turkey, Bulgaria, Iran

Fall leaves and fruit
The leaves have up to 12 pairs of veins and become burnished gold in fall. The fruit is a woody, bristly husk.

leaf up to 5 in (12 cm) long

finely pointed leaf bud

husk up to 1 in (2.5 cm) long

FEATURES

Bark Smooth, light gray; sometimes with shallow fissures in maturity

Leaves Dark, rich green upper surface

Fagus orientalis (Fagaceae)

ORIENTAL BEECH

This large, broadly spreading tree is in many ways similar to the common beech (p.152), with which it may hybridize on the boundary between their separate natural ranges. However, the oriental beech has larger leaves with more pairs of veins than the common beech, and can—given the right conditions—develop into an even larger tree.

The leaves are elliptic to obovate—at their widest above the middle. They have a wavy but untoothed margin and a pale green underside. Male and female flowers are small and yellow-green. However, while the male are drooping catkins, the female are erect inflorescences. They are borne in separate clusters on the same tree in spring. The flowers are followed by green, bristle-covered fruit husks, which become woody with age. Each husk contains up to three chestnut-brown, edible seeds.

♫ **Chinese beech (*Fagus lucida*)** This is a smaller tree than the oriental beech, up to 50 ft (15 m) tall, with smaller and narrower, glossy green leaves, 2–3 in (5–8 cm) long.

Siebold's leaves and fruit
The leaves are widest toward the base. The fruit is a ½in- (1.5 cm-) long, woody husk, held on a short, thick stalk.

bluntly toothed leaf margin

four-lobed husk covered in bristles

leaf up to 4 in (10 cm) long, 2 in (5 cm) wide

PROFILE

- ↕ 80–120 ft (25–35 m)
- ↔ Up to 70 ft (20 m)
- Deciduous
- Alternate simple
- Japan

Fagus crenata (Fagaceae)

SIEBOLD'S BEECH

Also known as the Japanese beech, this large, broadly spreading tree is native to Japan, but has been widely planted in the West since its introduction in 1892. It is named after the 19th-century German physician and naturalist Phillip Franz von Siebold. Siebold first identified the tree while working in Japan, where it is a dominant species in deciduous forests. The tree is one of the most popular species for bonsai (the art of growing miniature ornamental trees).

Siebold's beech is similar in outline to the common beech (p.152), with a large, rounded canopy. However, its leaves are more obovate in shape and the seed husks have flattened green whiskers at the base. Its bark is silver-gray and smooth, even in maturity. The leaves have 7 to 11 pairs of veins on the surface, and turn burnished gold in fall. Male flowers are borne in catkins, while tiny female flowers are carried in leaf axils. The fruit is a husk, which splits into four lobes to reveal up to three edible seeds.

leaf up to 4 in (10 cm) long, 2 in (5 cm) wide

PROFILE

- 70 ft (20 m)
- Up to 50 ft (15 m)
- Deciduous
- Alternate simple
- Japan

Tapering leaves
The leaves of the Japanese beech are bright green on the upper surface and ovate to elliptic, tapering abruptly at both ends.

parallel veins on leaf

wavy margin of leaf

FEATURES

Bark Silver-gray; remains smooth even in maturity

Fagus japonica (Fagaceae)

JAPANESE BEECH

Although the name of this tree may sometimes be confusingly given to Siebold's beech (p.155), the Japanese beech looks very different from *Fagus crenata*. This tree is much smaller in stature and often produces multiple stems, which gives it a shrublike appearance. It was introduced into western cultivation in 1907, but more than a hundred years later, it is still not a common tree and is rarely found outside botanical gardens and arboreta.

When they first unfurl from the leaf bud, the leaves are covered in dense white hairs, which may persist on the leaf underside. Each leaf has between 10 and 14 pairs of parallel leaf veins, all branching from a yellow-green midrib. In fall, the leaves turn yellow-orange and persist on the tree until the start of winter. Both male and female flowers are borne on the same tree, and female flowers are pollinated by pollen carried on the wind. The seed is an angular nut, contained within a bristly, woody husk, which opens to release the seed in fall.

PROFILE

⭥ Up to 80 ft (25 m)

⭤ Up to 50 ft (15 m)

Deciduous

Alternate simple

C. Chile, W. Argentina

_leaf up to 4 in (10 cm)
long, 2 in (5 cm) wide_

_ripe husk,
up to ½ in
(1 cm) long_

_oblong to ovate,
pendulous leaf_

Foliage and fruit
The leaves are distinguished from other
members of the _Nothofagus_ genus by the
14 to 18 pairs of parallel veins. The fruit
ripens from deep green to brown.

FEATURES

Bark Dark gray; becomes
cracked and fissured
in maturity

Nothofagus alpina (Fagaceae)

RAULI SOUTHERN BEECH

Earlier called _Nothofagus nervosa_ and _N. procera_,
this broadly conical tree is native to central Chile and
western Argentina, where it grows in the wild in the
foothills of the Andes. It is a fast-growing species that
produces excellent timber and is now planted in timber
plantations in the northern hemisphere.

The tree has distinctive upswept branches. The leaves,
at first glance, can be likened to those of the common
hornbeam (p.184). They are deep green, sometimes
bronze-green, on the upper surface and paler on the
underside, with some hairs on the midrib and the veins.
The shoots are green when young, becoming brown and
rough with age. The fruit, borne at the base of each leaf,
is a four-lobed husk containing three small nuts.

♀ **Oval-leaved southern beech (_Nothofagus betuloides_)** This
tree is distinguished by its smaller, more oval leaves, orange-brown
shoots, and a tighter, denser crown.

leaf up to 1 ¼ in (3 cm) long, ¾ in (2 cm) wide

unevenly toothed leaf margin

angular, green fruit husks in cluster

Leaves and fruit
Antarctic beech leaves are oblong to ovate, crinkled, and slightly cupped along the margins. The fruit is borne at the base of leaves on side shoots.

PROFILE

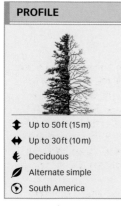

- ↕ Up to 50 ft (15 m)
- ↔ Up to 30 ft (10 m)
- ⚘ Deciduous
- ⬥ Alternate simple
- ⟳ South America

Nothofagus antarctica (Fagaceae)

ANTARCTIC BEECH

This fast-growing tree is broadly columnar and inhabits mountainsides from Cape Horn to northern Chile, where it is known by the common name *ñirre*. It was introduced into western cultivation in 1830.

The bark is dark gray-brown, developing deep fissures and cracking into irregular plates from an early age. The crown can be quite open, with branches that are sporadically large and sometimes twisted. The shoots are olive-gray on the underside, with a slight covering of hairs. Leaf buds are ovoid, shiny red-brown, although they sometimes have a purple bloom. Antarctic beech leaves have a deep glossy green upper surface and a paler underside, with four pairs of parallel veins. When crushed, they may emit a honey-like fragrance. In fall, they turn orange, red, and yellow before falling. The fruit is a four-winged husk with up to three seeds.

♧ **Myrtle beech (*Nothofagus cunninghamii*)** This evergreen beech has small, triangular leaves, which are up to ¾ in (2 cm) long. The leaves have tiny irregular teeth along the margins.

Roble beech foliage
The leaves are ovate to oval, sometimes almost oblong. They have 8 to 11 pairs of distinct veins.

green-brown husk

leaf up to 3in (7.5cm) long

irregularly toothed leaf margin

⬍ Up to 120ft (35m)

⬌ Up to 70ft (20m)

🌢 Deciduous

▱ Alternate simple

⊙ Chile, Argentina

FEATURES

Bark Gray and smooth when young; becoming cracked and fissured with age

Nothofagus obliqua (Fagaceae)

ROBLE BEECH

Also known as the coyan beech, the more common name for this tree is roble, which is the Spanish word for oak. In fact, in some respects, this large, broadly columnar tree does resemble the sessile oak (p.164). It produces tough, durable, oak-like timber, which in the past was used in shipbuilding and for making house beams and furniture. Today, it is widely planted as an ornamental tree, and grown alongside highways in some parts of southern Europe.

Its leaves are dark green on the upper surface and paler blue-green on the underside. Irregularly toothed along the margins, they have 8 to 11 pairs of distinct veins. In fall, they turn golden yellow before being shed. Small green flowers are produced at the base of leaves. Male flowers are borne singly; females, in threes. The fruit is a bristly husk containing up to three nuts.

◊ **Menzies' red beech (*Nothofagus menziesii*)** This evergreen tree is told apart by its shiny, purple-brown bark and double-toothed, small, orbicular leaves, up to ½in (1cm) long and wide.

PROFILE

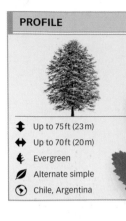

- ↕ Up to 75 ft (23 m)
- ↔ Up to 70 ft (20 m)
- ❧ Evergreen
- ⬤ Alternate simple
- ◉ Chile, Argentina

jagged serrations on leaf margin

stiff, oval to lanceolate leaf, up to 1–1½ in (2.5–4 cm) long

Dense foliage
Leaves are borne on shoots, which are red-brown above and green below. The leaves overlap so densely that, from a distance, the whole tree appears black in color.

fruit husk up to ¼ in (6 mm) long

FEATURES

Bark Dark slate-gray; smooth becoming widely fissured in maturity

Nothofagus dombeyi (Fagaceae)

COIGÜE SOUTHERN BEECH

Also called Dombey's southern beech, this broadly conical, evergreen tree is one of the most elegant of South American temperate trees. It was introduced into Europe in 1916. In this region, it may shed its leaves in winter, producing new ones in spring, with little evidence of a check in growth rate. In maturity, the tree resembles an old Atlas cedar (p.46), with an open arrangement of ascending branches, which bend outward toward the tips.

The bark flakes to reveal orange-red new bark. Borne on slender shoots, the leaves are glossy black-green on the upper surface, and matte, pale green on the underside. Male flowers are small, but bright red and contrast with the evergreen foliage. The fruit is a small, bristly husk, enclosing three small nuts.

◌ **New Zealand red beech (*Nothofagus fusca*)** Unlike the Coigüe southern beech, this tree's leaves are double-toothed with red veins on the underside, and turn dark red in fall.

PROFILE

coarsely toothed leaf margin

Spiky flowers
Male and female flowers are clustered in small, creamy yellow, spikelike catkins.

sharply pointed leaf tip

rounded leaf base

- ⬍ Up to 100 ft (30 m)
- ⬌ Up to 70 ft (20 m)
- 🍂 Deciduous
- 🌿 Alternate simple
- 🕐 Mediterranean, eastward into S.W. Asia

FEATURES

slender, sharp spines

Fruit Yellow-green husks

glossy red-brown seed

Seed Edible chestnut contained in a spiny husk

Castanea sativa (Fagaceae)

SWEET CHESTNUT

This fast-growing, broadly columnar, ornamental tree is indigenous to warm, temperate regions. It is widely cultivated elsewhere for its fruit—an edible chestnut. The Romans in particular valued the tree as a source of food and introduced it to many parts of their empire, including the British Isles.

The bark is smooth and light gray at first, developing spiral fissures in maturity. In winter, this characteristic distinguishes this tree from the pedunculate oak (p.165). The oblong leaves are glossy, dark green on the upper surface and paler on the underside. They are up to 8 in (20 cm) long and 3 in (7.5 cm) wide, with coarsely toothed margins. Each tooth links to a vein running to the midrib. Small, creamy yellow flowers are borne on erect catkins that ripen in midsummer. The fruit appears in fall, in a husk enclosing up to three chestnuts.

♤ **Japanese chestnut (***Castanea crenata***)** This tree is smaller than the sweet chestnut with long, narrow, bristle-toothed leaves, which appear more crinkly than its European counterpart.

spiky flower catkin, up
to 1 ½ in (4 cm) long

creamy yellow
female flower

golden yellow
underside of leaf

leaf up to 4 in (10 cm)
long, 1 in (2.5 cm) wide

green, spiny husk
containing seed

dark green uppe
surface of leaf

Leaves and fruit
The alternate leaves have a
leathery texture. Up to three nuts
are enveloped in a spiny husk,
which takes two years to ripen.

PROFILE

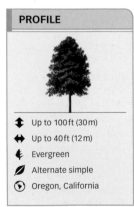

⬍ Up to 100 ft (30 m)

⬌ Up to 40 ft (12 m)

🌿 Evergreen

🍂 Alternate simple

◉ Oregon, California

Chrysolepis chrysophylla (Fagaceae)

GOLDEN CHESTNUT

This broadly conical, evergreen tree is often referred to
as the golden chinkapin and, sometimes botanically, as
Castanopsis chrysophylla. This is incorrect as the male
and female flowers on *Castanopsis* are borne on separat
catkins, whereas the male and female flowers on this
tree are on the same catkin. The name "golden" refers
to the underside of the leaf, which is covered with bright
golden hairs—an easy way of distinguishing it from othe
members of the Fagaceae family.

The gray bark is smooth when young but develops
fissures in maturity. The evergreen leaves are lanceolate
to oblong, broad in the middle, and taper to a point at each
end. Fragrant male and female flowers are borne on erect
catkins in summer. These are followed by dense clusters o
spiny husks, which contain three glossy brown nuts.

◌ **American sweet chestnut (*Castanea dentata*)** The
deciduous leaves are narrower and longer than those of the golden
chestnut. They are hairless, with saw-toothed leaf margins.

PROFILE

Up to 56 ft (17 m)

Up to 50 ft (15 m)

Evergreen

Alternate simple

W. United States

acorn ¾–1¼ in
(2–3 cm) long

Tanbark acorn
The fruit of the tanbark oak
is an ovoid acorn held within
a short, stalked, bristly cup.

hard, woody
exocarp (shell)

FEATURES

Leaves Oblong to elliptic,
sharply toothed, leathery

Lithocarpus densiflorus (Fagaceae)

TANBARK OAK

As the name suggests, this small to medium-sized,
pyramidal evergreen is closely related to oak (*Quercus*),
and displays some of the characteristics of this genus.
It is native to California and Oregon on the West Coast
and was introduced into European cultivation in 1874.
It has been regularly planted in parks, gardens, and
arboreta since then.

The bark is a valuable source of tannin, a chemical
used in the leather industry. It is gray-brown and smooth
at first, becoming vertically fissured in maturity. Both the
shoots and the leaves are covered with a soft, creamy
white down when young.

The leaves are similar to those of the sweet chestnut
(p.161) and are up to 4 in (10 cm) long and 2 in (5 cm) wide.
They are dark, glossy green on the upper surface, with
a paler green, sometimes almost white, underside. Fawn-
colored hairs surround the leaf veins. Clusters of small,
creamy yellow flowers are borne on erect spikes in late
spring and resemble the flowers of the sweet chestnut.

leaf up to 13 cm (5 in) long, 7.5 cm (3 in) wide

rounded, untoothed leaf lobe

glossy, dark green upper surface of leaf

PROFILE

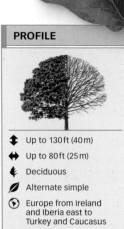

⬍ Up to 130 ft (40 m)

⬌ Up to 80 ft (25 m)

🍂 Deciduous

🌿 Alternate simple

🜂 Europe from Ireland and Iberia east to Turkey and Caucasus

Stalked foliage
This tree has elliptic leaves, which taper toward the base. Each leaf is attached to the twig by a yellow stalk, up to ½ in (1.25 cm) long.

FEATURES

Bark Light gray, smooth for the first 30 years; developing deep vertical fissures afterward

Quercus petraea (Fagaceae)

SESSILE OAK

Also known as the durmast oak, this long-lived, broadly spreading tree produces strong, durable timber, which is attractively figured and has been used over the centuries for shipbuilding, house construction, and for making furniture and wine barrels. In western Europe, the sessile oak often grows close to the pedunculate oak (p.165), and intermediate hybrids are regularly encountered.

The leaves have a dark green upper surface and a paler underside, with a slight covering of hairs. In late spring, the tree bears flowers in yellow-green, pendulous catkins. The seed is an acorn, one-third of which is enclosed in a scaly, unstalked cup, which is directly attached to the twig at its base.

⌂ Hungarian oak
This tree (*Quercus frainetto*) has obovate leaves, up to 8 in (20 cm) long, with large, deep, regular, oblong lobes.

Green acorn
The pedunculate oak's young fruit is a small green acorn, one-third of which is held in a cup at the end of a stalk.

small lobe at base of leaf

3–6 rounded lobes on either side of leaf margin

PROFILE

↕ 80–130 ft (25–40 m)
↔ Up to 100 ft (30 m)
↓ Deciduous
⬗ Alternate simple
⊙ Europe, W. Asia, N. Africa

acorn up to 1½ in (4 cm) long

FEATURES

Flowers Long, pendulous, greenish yellow catkins

Fruit Acorn, green ripening to brown in first year

Quercus robur (Fagaceae)

PEDUNCULATE OAK

Also known as English or common oak, this majestic, broadly spreading tree is found throughout Europe. It is usually found in lowlands, which allows its close relative sessile oak (p.164) to colonize upland regions of Europe. However, the two regularly hybridize where they meet, creating trees that may have characteristics of both species.

The pedunculate oak grows best on damp, rich, well-drained soils. It is long-lived, with many recorded veterans over 1,000 years old. The leaves are elliptic to obovate, tapering at the base. They are usually unstalked, or on shorter stalks than the sessile oak. The upper surface is deep green and the underside blue-green. The flowers are long, greenish yellow catkins. The fruit is borne on a stalk up to 4 in (10 cm) long.

◊ **Pyrenean oak**
This tree (*Quercus pyrenaica*) has longer leaves, which are slightly hairy on the upper surface.

leaf up to 5 in (12 cm) long, 3 in (7.5 cm) wide

elliptic to oblong leaf

acorn cup up to 1 in (2.5 cm) long

Foliage and fruit
Turkey oak leaves are elliptic to oblong and glossy dark green. The fruit are borne in cups covered in long, narrow scales.

PROFILE

- Up to 130 ft (40 m)
- Up to 80 ft (25 m)
- Deciduous
- Alternate simple
- C. and S. Europe

FEATURES

Bark Silver-gray, thick; deeply fissured in maturity

Quercus cerris (Fagaceae)

TURKEY OAK

This wide-spreading tree is found across central and southern Europe, from France through to Turkey. It normally has a long, straight trunk with relatively little branching on the first third of the trunk's height. Unlike other oaks, it does not yield valuable timber, because the timber cracks and splits when sawed. It has been widely cultivated across the rest of Europe since the 18th century, and has become naturalized in many areas. It is often planted as an ornamental tree.

The leaves are variably lobed, but usually with shallower and more lobes than the sessile oak (p.164) and the pedunculate oak (p.165). Both male and female flowers are yellow-green pendulous catkins, borne separately on the same tree in early summer. The fruit is an acorn, held in a cup covered with bristlelike scales for half of its length.

♤ **Chestnut-leaved oak (*Quercus castaneifolia*)** The leaves of this tree are oblong, tapering at both ends. They have coarse, triangular, sharply pointed teeth along the margins.

PROFILE

Leathery leaves
Leaves are elliptic to ovate, leathery, and occasionally toothed with pale gray hairs on the underside.

acorn up to ½ in (1.5 cm) long

dark green upper surface of leaf

sage green on leaf underside

leaf up to 3 in (7.5 cm) long, 2 in (5 cm) wide

↕ Up to 100 ft (30 m)

↔ Up to 100 ft (30 m)

❦ Evergreen

✎ Alternate simple

◉ S. Europe, Mediterranean region

FEATURES

Bark Gray-brown and smooth; develops shallow fissures and small plates with age

Flowers Male catkins are creamy gray

Quercus ilex (Fagaceae)

HOLM OAK

This evergreen oak is also known as holly oak or evergreen oak. It has a rounded, domed shape, and dense branching. The profusion of leaves on the branches is such that, from a distance, the canopy looks solid and black. Little light or rainwater penetrates through the canopy to ground level, so the ground beneath this tree is normally bare, except for dry, brown, spent leaves. The holm oak is native to southern Europe and the Mediterranean but has been cultivated in northern Europe since the 16th century.

It is a valuable tree in coastal regions, providing shelter against wind and salt spray. During cold winters, up to 50 percent of its leaves may be shed and replaced in spring. The male and female catkins appear separately on the same tree in summer. The fruit is a small, pointed acorn held in a fawn, scaly cup.

♫ **Japanese chinquapin (*Castanopsis cuspidata*)** This small, bushy tree is distinguished by its larger, lighter green leaves, which have a long drawn-out tip, and a yellow-green midrib and petiole.

tapering lobe
on leaf

leaf up to
8 in (20 cm) long

bristly lobe ends
in fine point

Deeply cut leaves
The alternate leaves are cut
into bristle-tipped lobes, and
are often borne on stalks that
are red at the base.

PROFILE

- Up to 100 ft (30 m)
- Up to 70 ft (20 m)
- Deciduous
- Alternate simple
- E. North America

FEATURES

acorn up to
1 ¼ in (3 cm)
long

shallow cup
holding acorn

Fruit Squat
acorn; ripening
in second year

Leaves Ruby-red mixed with
gold, before being shed in fall

Quercus rubra (Fagaceae)

RED OAK

This fast-growing, broadly spreading tree is native
to eastern North America from Alabama to Nova Scotia
and Quebec, Canada, where it grows farther north than
any other American oak. It is famed for its spectacular
leaf color displays in fall. Introduced into Britain as
early as 1724, the red oak has been widely planted
as an ornamental species in the temperate world
including western Europe.

It has a smooth, light gray bark, similar to beech when
young, which becomes deeply fissured in maturity. The
leaves are large, oval or obovate-shaped, with distinctive
lobes, each ending in a fine point. They have a dull matte
green upper surface, and pale blue-green underside,
with tufts of fine, brown hairs in the vein axils (where the
veins meet the midrib). The flowers are green-yellow,
pendulous catkins, which appear in spring.

♀ **Water oak (*Quercus nigra*)** This is a smaller, less vigorous tree
than the red oak. It has smaller, wedge-shaped leaves with variable
lobes, which are glossy, rich green on both sides.

Fall leaves
In fall, the scarlet oak leaves gradually change color, sometimes over four weeks, from green to vibrant scarlet before falling.

leaf 6 in (15 cm) long, 4 in (10 cm) wide

large, angular lobe on side of leaf

petiole up to 1½ in (4 cm) long

FEATURES

Leaves Dark, shiny green upper surface

fruit ¾ in (2 cm) long

Fruit Ovoid acorn, in shallow, scaled cup

Quercus coccinea (Fagaceae)

SCARLET OAK

Originating in the United States, the scarlet oak is a large, attractive, pyramid-shaped tree. It was introduced into cultivation in 1691. Today it is widely planted throughout the northern hemisphere in parks and gardens for the fall colors of its foliage.

The bark of the scarlet oak is silver-gray and smooth at first, but displaying warts and bumps. It matures to dark gray with shallow vertical fissures. The elliptic, occasionally oblong, leaves are shiny and pale green on the underside. They have three lobes on either side, each featuring spines toward the tip. Between the lobes, leaves may be cut almost to the midrib. The leaves are attached to red-brown shoots by thin, stalk-like petioles. The male flowers are borne in catkins at the tip of the previous year's shoots, while leaf axils bear the female flowers. The flowers of both sexes are pale green-brown.

🌢 **Pin oak (*Quercus palustris*)** The leaves of this tree are smaller than the scarlet oak, and are borne on longer petioles; and on the underside, carry tufts of brown hairs in the axils.

PROFILE

leaf shiny dark green on upper surface

⬍ Up to 70ft (20m)

⬌ 70ft (20m)

🍂 Evergreen

🌿 Alternate simple

⊙ W. Mediterranean, N. Africa, S.W. Europe

leaf up to 3in (7.5cm) long

Cork oak leaves
The leaves of this oak are variable, ranging from oval to oblong. Occasionally entire (smooth-edged), the leaves often have toothed margins.

broadly toothed leaf margin

FEATURES

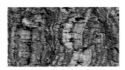

Bark Light brown, heavily fissured, corky

Fruit Slender, pointed acorn, up to 1¼ in (3cm) long

Quercus suber (Fagaceae)

CORK OAK

The broadly spreading, medium-sized cork oak is native to the western Mediterranean region, northern Africa, and the southern European Atlantic coast. It has been widely cultivated over many centuries for its thick, corky bark, which has been used to make cork products such as stoppers for wine bottles. Since the introduction of plastic corks and screw-top wine bottles, these trees have been neglected and felled right across the region but a move to return to cork stoppers may arrest the decline in this species.

Cork oak leaves are similar to those of the holm oak (p.167), with a pale green-gray, felt-covered underside. The male flowers are carried on weeping, yellow catkins in spring. The fruit is an acorn, green ripening to brown, and encased for half its length in a scaly cup.

◊ **Lucombe oak (***Quercus* x *hispanica* **'Lucombeana')** This cultivated form of the hybrid between the Turkey oak (p.166) and the cork oak is much larger than the cork oak.

PROFILE

leaf midrib

lobe ends in
sharp point

indentation
halfway to
midrib

dark gray-green
upper surface
of leaf

leaf bud

⬍ Up to 70 ft (20 m)

⬌ 50 ft (15 m)

🍂 Deciduous

🌿 Alternate simple

⊙ S. Europe from Spain
to Caucasus

...obed foliage
...he leaves of the downy oak vary
...om elliptic to obovate. The leaf
...obes are normally irregular in size.

FEATURES

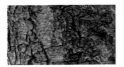

Bark Dark gray; becomes
fissured from an early age

Fruit Stalkless acorn, up
to 1½ in (4 cm) long; green
ripening to brown

Quercus pubescens (Fagaceae)

DOWNY OAK

This small to medium-sized, broadly spreading tree
has a large natural range, extending from Spain in the
west to the Caucasus in the east. The downy oak is
closely related to the sessile oak (p.164). Compared to
its cousin, the smaller downy oak has a more shrublike
appearance and ragged outline, and, as the name
suggests, its leaves have more hairs.

The variable, lobed leaves are pale green on the
underside with a dense covering of short, soft, gray-white
hairs. The leaf petioles (stalks) are also covered with hairs.
The leaf lobes may end in a small, sharp point, with a
few indentations between the lobes extending more
than halfway toward the leaf midrib. Clusters of male
flowers are borne on pendulous yellow-green catkins
in spring. The fruit is an acorn, contained in a hairy,
scaly cup for up to half of its length.

♤ **Lebanon oak (*Quercus libani*)** This tree has slender branches
and glossy green, oblong-lanceolate leaves. The leaf margins
have triangular, bristle-tipped teeth.

shiny, dark green
upper surface
of leaf

Algerian oak leaves
The leaves of this tree are
large, oblong-ovate to obovate,
sometimes oval. The largest lobes
appear where the leaf is broadest.

leaf up to 7 in
(19 cm) long, 4 in
(10 cm) wide

shallowly lobed
leaf margin

PROFILE

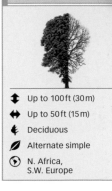

⬍ Up to 100 ft (30 m)

⬌ Up to 50 ft (15 m)

🍂 Deciduous

▱ Alternate simple

◉ N. Africa,
S.W. Europe

FEATURES

Bark Dark gray, almost black;
deeply fissured into small
square plates

Fruit Acorn, up to 1 ¼ in (3 cm)
long, carried in scaly cup

Quercus canariensis (Fagaceae)

ALGERIAN OAK

Also known as Mirbeck's oak, this tree is native
to northern Africa and southwestern Europe including
the Canary Islands. It is reasonably hardy and survives
much further north, where it is planted in parks, gardens,
and arboreta. In maturity, the
Algerian oak has a dense rounded
head of branches on top of a
straight trunk. Although not
strictly evergreen, many of its
leaves remain green and stay
on the tree through winter.

Each leaf has between 6
and 13 pairs of lobes, with a
matte, slightly blue-green, hairy
underside. The male flowers are
bunched at leaf nodes in yellow
pendulous catkins. The fruit
is attached to the shoot by a
½ in- (1 cm-) long stalk.

Young oak
The Algerian oak is a large,
narrowly columnar tree
when young, broadening
out in maturity.

PROFILE

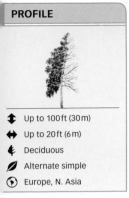

↕ Up to 100 ft (30 m)

↔ Up to 20 ft (6 m)

🍂 Deciduous

🌿 Alternate simple

⊙ Europe, N. Asia

red-brown leaf stalk

leaf up to 2½ in (6 cm) long, 1½ in (4 cm) wide

Catkins and fruit
Erect female flowers develop into pendulous, cucumber-shaped fruit in fall. Leaves are dark green on the upper surface and pale green on the underside.

fruiting catkin 1¼ in (3 cm) long

triangular leaf tapers to a point

FEATURES

Bark White or silver-gray; becomes corky with black fissures in old age

---

Betula pendula (Betulaceae)

SILVER BIRCH

The natural range of the silver birch extends from Britain and Ireland, across Europe and northern Asia, to the Pacific Ocean. This narrowly weeping tree seldom reaches its full height potential, and is conical with upswept branches when young. It is hardy, being able to withstand intense cold and long periods of drought, and it seeds prolifically to establish itself on cleared ground. Also known as the warty birch, it has shoots covered in wart-like bumps, making them rough to the touch.

The leaves are borne on slender shoots and turn butter yellow in fall. Both male and female flowers appear separately on the same tree in spring. The male flowers are borne in pendulous clusters. The green fruit ripen to brown and break up on the tree to release small, winged seeds.

🌢 **Gray birch (*Betula populifolia*)** This American counterpart of the silver birch is distinguished by its poplar-like leaves, which are borne on a long lax stalk and flutter in the slightest breeze.

PROFILE

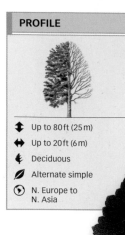

- ↕ Up to 80 ft (25 m)
- ↔ Up to 20 ft (6 m)
- Deciduous
- Alternate simple
- N. Europe to N. Asia

leaf up to 2 in (5 cm) long and wide

ovate to rounded leaf

upright fruiting female catkin

Leaves and fruiting catkins
The dark green leaves have toothed margins and taper to a point. The green, fruiting female catkin ripens and breaks off from the tree.

FEATURES

Bark Gray when young, matures to white during later years

male catkin up to 4 in (10 cm) long

Male flowers Yellow, pendulous catkins; release fine, yellow pollen particles in spring

Betula pubescens (Betulaceae)

DOWNY BIRCH

Also known as white birch, this hardy, broadly conical, deciduous tree has a natural range that extends across northern Europe into northern Asia. It thrives in poor soils and wet conditions, and grows right up into the Arctic region. It is called "downy" because of its soft, hairy shoots, a feature that distinguishes it from the silver birch (p.173), which has rough shoots. Other characteristics distinguishing these two birches include the less pendulous form and twisting branches of the downy birch.

While the bark on the tree's trunk is gray-white, its shoots and branches are a glossy, copper brown. The leaves are dark green on the upper surface and paler green on the underside, with hairy veins. Unlike the male flowers, which are drooping catkins, the female flowers are upright, green catkins. They are borne separately from the male flowers on the same tree in spring. They turn brown when ripe and shed large amounts of small, papery, winged seeds.

drooping male catkin, borne at end of shoot

- ⬍ Up to 80ft (25m)
- ⬌ Up to 20ft (6m)
- 🍂 Deciduous
- 🍃 Alternate simple
- ⊙ W. China

upright female catkin

ovate leaf, up to 3in (7.5cm) long, 2in (5cm) wide

saw-toothed leaf margin, with each tooth pointing forward

Long-lasting catkins
Male and female catkins emerge on the same plant in summer and open the following spring.

FEATURES

Bark Peels in thin strips, cream-pink when exposed

fruiting female catkin

Female flowers Green catkins, borne upright

Betula albosinensis (Betulaceae)

CHINESE RED-BARKED BIRCH

This deciduous tree is best known for its coppery to orange-red bark, which peels to reveal a fresh cream-pink surface beneath. This happens on the branches as well as the trunk. The tree is native to western China, where it grows high in mixed woodland in mountainous regions. It was introduced in Europe in 1901 by English plant collector Ernest H. Wilson, and has been planted around the temperate world ever since.

In spring, leaves emerge from small, sticky, red buds, and are covered in white down. This fades to reveal glossy mid-green leaves that turn golden yellow in fall, before being shed. Male and female flowers, carried in catkins, appear separately on the same tree. Female catkins break up to release hundreds of winged seeds.

♤ **Chinese dwarf birch (*Betula chinensis*)** Apart from being much smaller tree, with smaller sharply toothed leaves, this bir also lacks the striking bark of the Chinese red-barked birch.

PROFILE

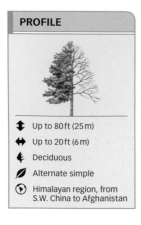

⭥ Up to 80 ft (25 m)

⭤ Up to 20 ft (6 m)

🌢 Deciduous

🗡 Alternate simple

⊙ Himalayan region, from S.W. China to Afghanistan

leaf up to 4 in (10 cm) long, 2½ in (6 cm) wide

serrated margin

teardrop-shaped leaf tapers to a point

green, upright female catkin

each vein ends in a sharp tooth

Fruiting catkins
The female catkins of the Himalayan birch ripen from green to brown in spring, and disintegrate on the tree to release small, winged seeds.

FEATURES

Bark Paper thin, peels off in horizontal strips

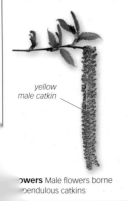

yellow male catkin

...owers Male flowers borne ...pendulous catkins

Betula utilis (Betulaceae)

HIMALAYAN BIRCH

This attractive birch has a wide natural distribution throughout the Himalayan region, from Afghanistan in the west to southwestern China in the east. It is a broadly conical, medium-sized deciduous tree. It occasionally reaches 80 ft (25 m) in height, but is usually smaller.

Within the wild population, the bark color is extremely variable, ranging from pink-brown or orange-red, through to white. However, the orange-red form is most common in cultivation. The bark peels off in long, ribbonlike strips. The leaves are glossy dark green on the upper surface, and paler on the underside, with some hairs on the midrib. Each ovate leaf is distinctly veined, with up to 12 pairs of parallel veins. It is attached to hairy shoots by a short leaf stalk ½ in (1.25 cm) long. Pendulous male and upright female flowers are borne on the same tree in catkins.

♤ **Monarch birch (*Betula maximowicziana*)** A bigger tree than the Himalayan birch, this related species can be easily distinguished by its much larger heart-shaped leaves.

PROFILE

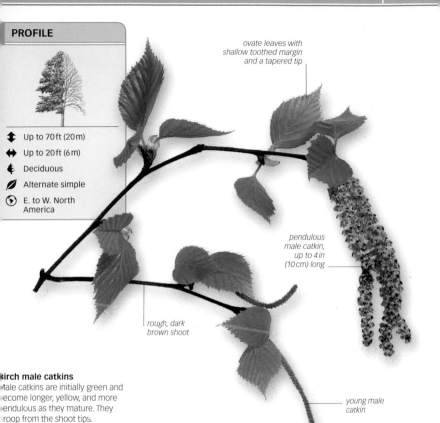

▲ Up to 70 ft (20 m)

↔ Up to 20 ft (6 m)

🍂 Deciduous

🍃 Alternate simple

⊙ E. to W. North America

ovate leaves with shallow toothed margin and a tapered tip

pendulous male catkin, up to 4 in (10 cm) long

rough, dark brown shoot

young male catkin

Birch male catkins
Male catkins are initially green and become longer, yellow, and more pendulous as they mature. They droop from the shoot tips.

FEATURES

Bark Creamy white; peels in thin papery strips

Leaves Dark green to orange, before being shed in fall

Betula papyrifera (Betulaceae)

PAPER-BARK BIRCH

This broadly conical, medium-sized deciduous birch has a natural range that extends across North America, from the Atlantic to the Pacific coast. Native Americans used the tough, durable bark of this tree to cover their canoes and make them waterproof. As a result, the tree is often referred to as canoe birch. A hardy tree, it grows as far north as Alaska and Labrador, and was introduced into cultivation in Europe around 1750. It has been a popular addition to parks and gardens ever since.

The leaves are dark green and smooth on the upper surface, pale green on the underside, with veins lightly covered in fine hairs. In fall, the leaves turn marmalade-orange before falling. The flowers are borne in catkins. Female catkins are up to 3 in (7.5 cm) long, ripen from green to brown, and disintegrate on the tree to release small, brown, winged seeds.

♀ **Japanese cherry birch (*Betula grossa*)** This tree is distinguished by its dark bark, hornbeam-like leaves, and a unique fragrance, given off when its twigs are broken.

PROFILE

- Up to 70 ft (20 m)
- Up to 25 ft (8 m)
- Deciduous
- Alternate simple
- Russia, Japan, Korea

Rough shoots
The leaves and catkins of the Russian rock birch are carried on coarse, warty shoots.

deeply serrated leaf margin

upright fruiting female catkin

leaf margin ends in a point

rough-textured shoot

leaf up to 3½ in (9 cm) long, 2½ in (6 cm) wide

drooping, yellow male catkin

FEATURES

Bark Cream at first, peels to show pink or pink-fawn patches

erect, green catkin

Female flowers Upright, smaller than male catkins; ripening to brown

Betula ermanii (Betulaceae)

RUSSIAN ROCK BIRCH

This medium-sized, long-lived birch is native to Japan and Korea, as well as parts of Russia. Sometimes called Erman's birch, it was introduced into cultivation in 1890. It is a broadly conical tree, and one of relatively few birches to grow with multiple stems. A distinctive feature of this tree is the way its peeled bark shreds and hangs like a ragged fringe along the undersides of branches.

The bark is creamy, with a tint of pink, and stands out particularly when the tree is bare in winter. The leaves are glossy dark green on the upper surface and paler on the underside. They are triangular to ovate, with 7 to 11 pairs of impressed, parallel veins, and fawn-colored hairs in the vein axils. Both male and female flowers are 1½ in- (4 cm-) long catkins. They are carried in separate clusters on the same tree in spring. Female catkins break up on the tree to release small, winged seeds.

♀ **Korean birch (*Betula costata*)** Though closely related to the Russian rock birch, this tree has leaves that are less heart-shaped at the base and bark that has a greater pink tinge.

PROFILE

- 50–100 ft (15–30 m)
- Up to 50 ft (15 m)
- Deciduous
- Alternate simple
- C. and E. United States

Drooping male catkins
Male flowers are borne in pendulous clusters in spring before the leaves unfurl. Female flowers appear on the same tree but in separate catkins.

yellow male catkin, up to 3 in (7.5 cm) long

FEATURES

Bark Peels in fine papery strips, making the tree appear shaggy

sharply toothed leaf margin

Leaves Ovate; up to 4 in (10 cm) long

Betula nigra (Betulaceae)

RIVER BIRCH

Also known as red birch, this broadly spreading tree reaches 100 ft (30 m) in the wild but is smaller in cultivation. It occurs naturally from Maine in the east to Minnesota in the central United States, and south to Texas, where, as the name suggests, it grows beside creeks and in low-lying marsh ground.

It has attractive pink-brown bark when young, turning purple-black with striking orange fissures with age. This feature has made it a popular ornamental species. The leaves have a deep green upper surface and a blue-green underside with silvery hairs on the veins. The flowers are borne in catkins. Female catkins are upright and green, males are pendulous.

⚲ Yellow birch
This tree (*Betula alleghaniensis*) has a shining yellow bark and more finely toothed leaf margins.

up to 10 pairs of distinct, opposite leaf veins

green, unripe fruit

leaf up to 4 in (10 cm) long, 3 in (8 cm) wide

Circular leaves
The European alder's leaves are obovate to circular, finely toothed, with a strongly indented midrib.

FEATURES

cone up to ¾ in (2 cm) long

Fruit Brown woody cone, with scales opening to release numerous seeds

Alnus glutinosa (Betulaceae)

EUROPEAN ALDER

This broadly conical tree is native to Europe, western Asia, and North Africa. It is normally found growing in damp soil, often close to rivers and lakes. When its roots are submerged in water for long periods of time, it is able to create its own oxygen supply. Alder timber is water-durable and has been used to make boats, footwear, and the timber piles upon which Venice is built.

The tree has dark, gray-brown bark, which is lightly fissured from an early age. Its leaves are a shiny dark green on the upper surface and pale gray-green on the underside, with tufts of hairs in the leaf axils. Both male and female flowers are catkins. The male is greenish yellow, pendulous, and up to 4 in (10 cm) long. The female is upright, red, and oval-shaped, ripening into a small brown cone after fertilization.

✿ **Oriental alder (*Alnus orientalis*)** This medium-sized alder is easily distinguished from the European alder by its sticky winter buds and its ovate, coarsely toothed, glossy leaves.

PROFILE

- Up to 70 ft (20 m)
- Up to 40 ft (12 m)
- Deciduous
- Alternate simple
- Europe

leaf underside has grey hairs

leaf up to 4 in (10 cm) long, 2 in (5 cm) wide

fruit contains small, winged seeds

Leaves and cone-like fruit
The green fruit ripen to become brown, woody, cone-like structures. Leaves are matte dark green on the upper surface.

FEATURES

catkin 4 in (10 cm) long

Male flowers Orange-yellow; borne on pendulous catkins

Alnus incana (Betulaceae)

GRAY ALDER

Native to most of Europe except for the British Isles, the gray alder is a fast-growing, medium-sized, broadly conical tree. It grows at elevations of up to around 3,280 ft (1,000 m) in the Caucasus mountains. This hardy species is regularly planted in cold, wet conditions where other trees may not thrive. It is also used to cover land being reclaimed after industrial activities such as mining and landfill. The tree gets its name from the dense covering of short gray hairs on the underside of the leaves.

The bark is dark gray and smooth, even in maturity. Leaves have an irregular, double-toothed margin, ending in a pointed tip. Flowers are borne clustered in catkins. Male and female flowers bloom separately on the same tree in early spring. The female catkin is erect and red. The fruit releases seeds after ripening, and remains on the tree for a long time.

◊ **Sitka alder (*Alnus sinuata*)** Unlike the gray alder, this tree does not have a dense, hairy covering on the underside of the leaf. It is also much smaller and more shrublike in form.

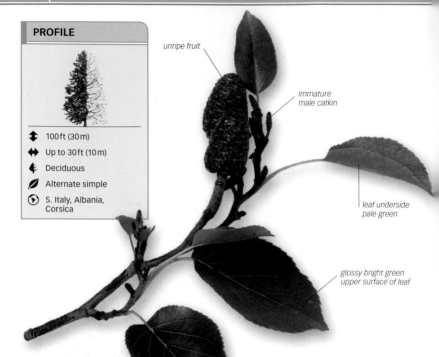

unripe fruit

immature male catkin

leaf underside pale green

glossy bright green upper surface of leaf

PROFILE

↕ 100 ft (30 m)

↔ Up to 30 ft (10 m)

🍂 Deciduous

🌿 Alternate simple

⊙ S. Italy, Albania, Corsica

broadly heart-shaped leaf base

Bright green foliage
Heart-shaped, occasionally almost round leaves are widely spaced on the shoots. Female catkins turn from pale brown to red.

FEATURES

catkin 3 in (7.5 cm) long

Male flowers
Long, drooping, yellow catkins

Fruit Ripen to brown, woody cone-like structures

Alnus cordata (Betulaceae)

ITALIAN ALDER

The broadly conical Italian alder is the largest species of its genus. It is native to southern Italy, parts of Albania, and Corsica. It grows at elevations of up to 3,280 ft (1,000 m) above sea level on dry mountain slopes. This is unusual for a species of alder, as most require moist soil conditions to thrive. This tree takes its species name *cordata* from the fact that it produces cordate, or heart-shaped, leaves.

The bark is dull gray tinged with green, smooth at first, becoming fissured with age. Branches in the crown of the tree tend to have lighter bark. The leaves are up to 4 in (10 cm) long, with a sparse covering of hairs on the underside. They are sharply toothed around the margin except for the base of the leaf, which may be entire (smooth). Both male and female flowers appear separately on the same tree in early spring.

♧ **Japanese alder (*Alnus japonica*)** This related tree has narrower, more pointed, elliptic leaves, and bears smaller male catkins in late winter or early spring.

male flower

drooping male catkin, up to 6 in (15 cm) long

- 50 ft (15 m)
- Up to 30 ft (10 m)
- Deciduous
- Alternate simple
- W. North America

Pendulous male catkins
The red alder flowers in early spring. Male flowers are carried in pendulous catkins, ripening to orange.

FEATURES

Leaves Up to 4 in (10 cm) long, 3 in (7.5 cm) wide, with slightly pointed tips

fruit up to 1 in (2.5 cm) long

Fruit
Egg-shaped, woody brown; containing many winged seeds

Alnus rubra (Betulaceae)

RED ALDER

Also known as the Oregon alder, this fast-growing tree is native to the North American west coast, from Alaska to California. It is prominent in the Cascade valleys and Coast Mountain areas, where it grows on areas of disturbed soil and along highways. It thrives in damp soil conditions, especially alongside creeks and rivers.

In the wild, the bark is normally silver-white and resembles white-barked birches; however, in cultivation the bark is usually relatively smooth and pewter-colored with a pinkish tinge. The leaves have a dark green and smooth upper surface with reddish veining, and a pale gray-green underside with rusty red hairs along leaf veins. They are attached to thin red shoots by red petioles up to 1½ in (3.5 cm) long. Male and female flowers are borne separately on the same tree. The smaller, bright red female flowers ripen into cone-like fruit.

♀ **Japanese green alder (*Alnus firma*)** Unlike the red alder, this smaller, shrublike tree has green petioles and no rust-colored hairs on the underside of the leaf.

PROFILE

- Up to 80 ft (25 m)
- 50 ft (15 m)
- Deciduous
- Alternate simple
- S.E. Europe, W. Asia

irregular serrations around leaf margin

Unripe fruit and leaf
The leaves are broadly oval but heart-shaped at the base. The fruit when unripe is a green husk.

leaf up to 4 in (10 cm) wide

fruit covered in long bristles

FEATURES

Bark Light gray-brown; turns distinctly corky in maturity

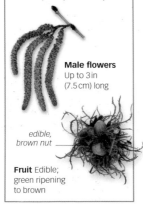

Male flowers
Up to 3 in (7.5 cm) long

edible, brown nut

Fruit Edible; green ripening to brown

Corylus colurna (Betulaceae)

TURKISH HAZEL

This magnificent tree is native to southeastern Europe and western Asia. It was introduced into cultivation in western Europe, including Britain, in the 16th century, and has been widely planted in parks and gardens since then. With its symmetrical, broadly conical form, it is ideally suited as an avenue or street tree.

The leaves have a dark green upper surface and are paler on the underside. Each leaf is irregularly serrated, and can grow up to 6 in (15 cm) long. In fall, they turn a rich yellow before being shed. Male and female flowers are carried in separate catkins on the same tree. Pendulous male flowers are yellow, while the smaller female catkins are red.

Stately form
In cultivation, the Turkish hazel normally has a short, straight trunk with a pyramid-shaped canopy.

Pendulous fruit and leaves
The fruit forms in hop-like clusters, with each husk containing a nutlike seed. The thick, stiff leaves have 12 to 15 pairs of veins.

dark, glossy green upper surface of leaf

ovate leaf, 3–5 in (8–13 cm) long

double-toothed leaf margin

bladderlike, flat husk

fruit up to 2 in (5 cm) long

PROFILE

- Up to 70 ft (20 m)
- 50 ft (15 m)
- Deciduous
- Alternate simple
- S. Europe, W. Asia

FEATURES

Bark Gray-brown; fissures into irregular loose flakes

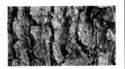

male flower up to 4 in (10 cm) long

Flowers Male and female borne in catkins on the same tree

Ostrya carpinifolia (Betulaceae)

HOP HORNBEAM

This round-headed, broadly spreading tree is native to southern Europe and western Asia. The name *Ostrya* is derived from the Greek word *ostrua*, which means "bone-like," referring to its hard, heavy timber. It was introduced into cultivation in 1724, and is admired for its fall colors. Occasionally, the hop hornbeam produces long horizontal branches, which seem to defy gravity before they eventually need to be propped up.

Smooth at first, the bark cracks in flakes, which fall to reveal orange patches beneath. The leaves resemble those of the common hornbeam (p.184), and are pale green on the underside, with hairs in the vein axils (where the veins meet the midrib). They turn a clear bright yellow before being shed in fall.

◇ Ironwood
This smaller tree (*Ostrya virginiana*) has hairy shoots and leaves with fewer veins.

Leaves and flowers
Fragrant flowers are borne in clusters of up to 10 in early to midsummer. The leaves vary from heart-shaped to almost rounded.

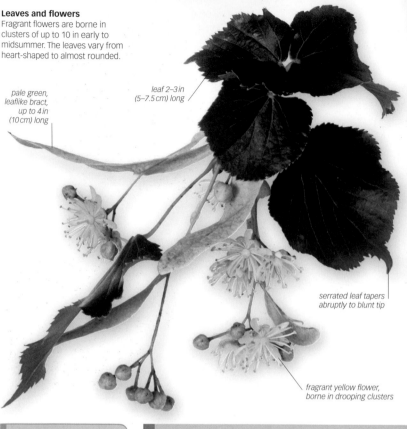

pale green, leaflike bract, up to 4 in (10 cm) long

leaf 2–3 in (5–7.5 cm) long

serrated leaf tapers abruptly to blunt tip

fragrant yellow flower, borne in drooping clusters

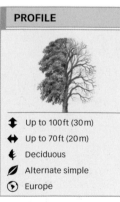
Tilia cordata (Malvaceae)

SMALL-LEAVED LIME

Endemic to Europe, from Portugal east to the Caucasus, this broadly columnar tree is known to have been regularly coppiced (trimmed back to stumps) in the past. When found growing wild in woodland, it is a good indicator that the woodland is old or even ancient. Some coppiced trees are believed to be over 2,000 years old.

The bark is silver-gray and smooth, becoming fissured from around 40 years of age. The inner bark, known as bast, is very fibrous and was often woven to make rope. Cordate, or heart-shaped, leaves are glossy green on the upper surface and blue-green on the underside, with some hairs in the vein axils. They turn yellow in fall before being shed. The flowers are accompanied by a narrow bract and develop into rounded fruits, up to ½ in (1 cm) wide. Each fruit has a green-gray feltlike surface and clusters are attached to the tree by a long stalk.

♤ **Crimean lime (*Tilia x euchlora*)** This tree is easily distinguished from the small-leaved lime by its smaller size, pendulous lower branches, and glossy, deep green leaves.

leaf up to 6 in
(15 cm) wide

pale green,
downy floral bract ___

pale yellow
flower ___

Sharply-toothed leaves
Rounded to broadly oval, the
leaves are sharply toothed
around the margin.

PROFILE

↕ Up to 100 ft (30 m)

↔ Up to 50 ft (15 m)

🍂 Deciduous

🌿 Alternate simple

⊕ Europe, S.W. Asia

Tilia platyphyllos (Malvaceae)

LARGE-LEAVED LIME

Also known as the broad-leaved lime, this large,
broadly columnar ornamental tree has a round-domed
canopy, a long, straight trunk, and graceful, arching
branches. It is native to Europe and southwest Asia,
and is commonly planted in parks and along avenues
throughout this region.

The smooth, light gray bark develops some shallow
fissures in maturity. Its leaves, large for this genus, are
slightly hairy and deep green on the upper surface and
light green on the underside, with dense hairs along the
midrib and in the vein axils. The leaf stalks are covered
in soft white hairs. Fragrant flowers are produced in
pendulous clusters of up to six in early summer. A bract,
up to 5 in (12.5 cm) long and ½ in (1.5 cm) wide, occurs
alongside each flower cluster. The fruit is a pale green,
downy nutlet borne on a long stalk in late summer.

🌷 **Japanese lime (*Tilia japonica*)** This tree has smaller leaves up
to 3 in (8 cm) long. They are narrower, more heart-shaped at the base,
and have a longer drawn-out tip.

Foliage and flowers
The leaves have sharply toothed margins and pointed tips. Fragrant flowers are borne in pendulous clusters of up to 10.

broadly ovate leaf, heart-shaped at base

leaf up to 4 in (10 cm) long and wide

yellow flower borne in small, drooping cluster

pale green, linear flower bract

PROFILE

⬍ 100–130 ft (30–40 m)

⬌ Up to 50 ft (15 m)

🍂 Deciduous

🌿 Alternate simple

📍 Europe

FEATURES

Fruit Pealike, light green; emerge in mid- to late summer

Tilia x europaea (Malvaceae)

COMMON LIME

Also known as the European lime, this large, broadly columnar tree is a hybrid between the small-leaved lime (p.188) and the large-leaved lime (p.189). It is found naturally wherever the ranges of the two parents overlap—which is across most of Europe. The tree does not have the elegance of either parent and produces unsightly suckers around the base of its trunk. It is also prone to aphid attack, which results in a coating of sticky aphid excrement known as honeydew.

The bark is gray to gray-brown, smooth when young, developing shallow, vertical fissures in maturity. The leaves are dull dark green on the upper surface and pale green on the underside, with tufts of hairs in the main vein axils (where the veins meet the midrib). The flowers, carried alongside a pale green bract in early summer, are followed by rounded fruit.

◊ **Henry's lime (*Tilia henryana*)** This slow-growing Chinese lime has large leaves, which have prominent bristlelike teeth along the margins.

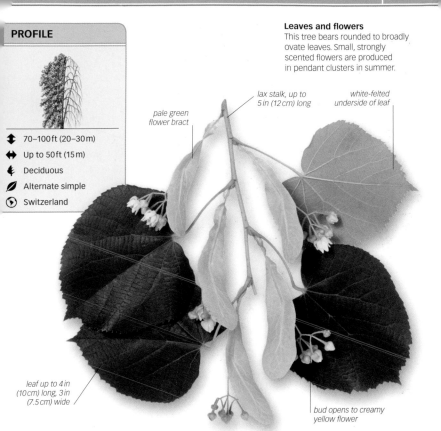

Leaves and flowers
This tree bears rounded to broadly ovate leaves. Small, strongly scented flowers are produced in pendant clusters in summer.

lax stalk, up to 5 in (12 cm) long

white-felted underside of leaf

pale green flower bract

leaf up to 4 in (10 cm) long, 3 in (7.5 cm) wide

bud opens to creamy yellow flower

↕ 70–100 ft (20–30 m)
↔ Up to 50 ft (15 m)
🍂 Deciduous
▱ Alternate simple
◉ Switzerland

FEATURES

Fruit Rounded, light green, and pealike

Tilia tomentosa 'Petiolaris' (Malvaceae)

WEEPING SILVER LIME

One of the most ornamental of large trees, the weeping silver lime has a broadly columnar habit. Its long, arching branches are often likened to the vaulted ceiling of a cathedral. The weeping silver lime is a cultivar of silver lime (*Tilia tomentosa*), and is believed to have been first raised in Switzerland in the mid-1800s.

The bark is slate-gray, smooth when young, developing vertical fissures in maturity. Quite often, a graft union is evident as a slight swelling on the trunk between 6 ft (2 m) and 15 ft (5 m) from the ground. Weeping shoots hang from arching branches and carry leaves that have a dark green upper surface and a white-felted underside. The leaves flash a distinctive white and green as they flutter in the breeze.

◔ **Oliver's lime**
This tree (*Tilia oliveri*) has a less weeping ha hairless shoots, and silver-gray bark.

Leaves and flowers
The American basswood has large leaves, more than 10 in (25 cm) long and 8 in (20 cm) wide. Fragrant flowers bloom in early summer in pendulous clusters.

forward-pointing teeth along leaf margin

deep green leaf upper surface

leaf tapers to pointed tip

pale yellow flower

oblong bract accompanies flower cluster

PROFILE

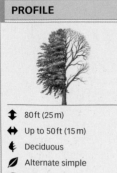

- 80 ft (25 m)
- Up to 50 ft (15 m)
- Deciduous
- Alternate simple
- E. North America

Tilia americana (Malvaceae)

AMERICAN BASSWOOD

Also known as American lime, this tree is native to eastern North America, from New Brunswick, Canada, south to Tennessee and Arkansas. "Bass" refers to the inner bark, which is extremely fibrous. Early European settlers would peel off strips of the bark and weave it into a coarse rope. The tree is cultivated both within and outside North America, and was introduced into Europe in 1752.

The bark is gray-brown and smooth at first, becoming rough and corky, with long scaly ridges in maturity. The leaves, which feel slightly rough to the touch, are broadly ovate, sometimes almost rounded, with irregularly serrated leaf margins. Slightly shiny on the upper surface, the leaves are pale green underneath, with some tufts of hair in the vein axils. In winter, buds are a distinctive light green color. The fruit is a woody, rounded drupe, covered in gray-brown hairs.

♀ **Mongolian lime (*Tilia mongolica*)** Apart from its more compact, rounded habit and smaller size, this tree also has smaller leaves, up to 3 in (7.5 cm) long, carried on red stalks.

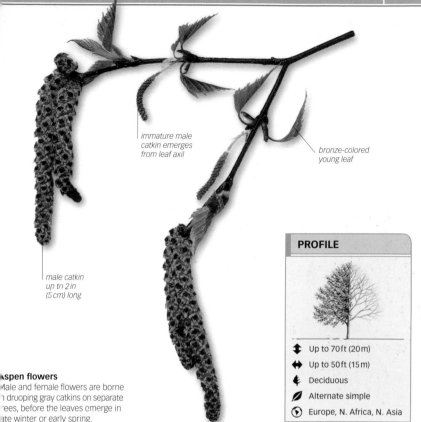

immature male catkin emerges from leaf axil

bronze-colored young leaf

male catkin up to 2 in (5 cm) long

Aspen flowers
Male and female flowers are borne in drooping gray catkins on separate trees, before the leaves emerge in late winter or early spring.

PROFILE

Up to 70 ft (20 m)

Up to 50 ft (15 m)

Deciduous

Alternate simple

Europe, N. Africa, N. Asia

FEATURES

sharply toothed leaf margin

leaf up to 3 in (7.5 cm) long and wide

Leaves Broadly ovate to rounded, three-veined; borne on slender, flattened leaf stalks

Populus tremula (Salicaceae)

ASPEN

Also known as the European aspen, this medium-sized deciduous tree is native to the whole of Europe and northern Asia, and is also indigenous to parts of North Africa. Its leaves tremble and quiver in the slightest breeze, hence its species name *tremula*.

Aspen is conical when young, becoming broadly spreading with age. It has light and dark gray, smooth bark, which becomes ridged at the base in maturity. Young leaves emerge in spring, fading to dull green with three prominent yellow-green veins at the base. They have a dark green upper surface and a pale gray-green underside. In late spring, tiny seeds covered in white, cottony hairs are released from green, capsule-like fruit. The hairs on the seeds help carry them in the air for some distance.

♤ **Balsam poplar (*Populus balsamifera*)** An erect-branched habit and a shorter leaf petiole distinguish this tree from the aspen. In spring, the leaves emit a strong balsam-like fragrance.

ovate to triangular leaf

slender, flattened petiole (leaf stalk)

PROFILE

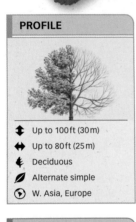

↕ Up to 100 ft (30 m)

↔ Up to 80 ft (25 m)

🍂 Deciduous

🌿 Alternate simple

◉ W. Asia, Europe

Glossy leaves
Bronze-orange when new, leaves turn deep green on the upper surface. On the underside, they are pale sage-green with silver-gray down.

leaf up to 3 in (8 cm) long and wide

FEATURES

Bark Pale gray, dark, diamond-shaped pattern; develops large burrs with age

male catkin up to 4 in (10 cm) long

Male flowers Gray with red anthers; borne in drooping catkins in early spring

Populus nigra (Salicaceae)

BLACK POPLAR

This large, deciduous tree is native to the whole of western Asia and Europe. It is a broadly spreading tree with long heavy branches. The black poplar inhabits river valleys and damp areas throughout its natural range, but in western Europe, including parts of the British Isles, it gives way to the Atlantic subspecies *P. nigra* subsp. *betulifolia,* which has downy shoots and young leaves.

The leaves emerge from large, conical, shiny buds, which smell of balsam when crushed. Male and female flowers are borne on separate trees. In late spring, tiny seeds, covered in white, cotton- wool-like hairs, are released from green fruit. They may be carried in the air for some distance.

♧ Hybrid poplar
This tree (*P.* x *canadensis*) bears lighter branches and grows faster than the black poplar.

ovate leaves
The Lombardy poplar bears triangular to diamond-shaped leaves, carried on a flattened, yellow-green petiole (leaf stalk), up to 1 in (2.5 cm) long.

pale green leaf underside

leaf up to 2½ in (6 cm) long

PROFILE

↕ Up to 100 ft (30 m)
↔ Up to 80 ft (25 m)
🍂 Deciduous
🌿 Alternate simple
◑ Europe, W. Asia

Populus nigra 'Italica' (Salicaceae)

LOMBARDY POPLAR

This narrowly columnar tree is one of the most easily recognizable of all trees because of its slender, column-like outline and strongly ascending branching. It is not a true species but a distinctive male variety of the black poplar (p.194). This tree is believed to have naturally occurred in northern Italy in the early 1700s. Since then it has been widely propagated from cuttings, and is now one of the most common trees planted throughout the temperate world, especially in western Europe.

The bark is dark gray, with fluting and buttressing toward the trunk's base. The leaves are glossy bright green on the upper surface, with regular small, curved teeth along the leaf margins. Male catkins are up to 3 in (7.5 cm) long, deep red, turning red and yellow when shedding pollen in mid-spring. This tree does not produce female flowers.

♤ **Chinese balsam poplar (*Populus szechuanica* var. *tibetica*)**
This poplar has a broadly columnar habit and large, broadly ovate leaves, which have a crimson midrib.

Heart-shaped leaves
The broadly ovate to heart-shaped leathery leaves have conspicuous red leaf stalks.

leaf up to 14 in (35 cm) long, 8 in (20 cm) wide

↕ Up to 70 ft (20 m)

↔ Up to 30 ft (10 m)

🍂 Deciduous

🍃 Alternate simple

⊙ C. China

bright green upper surface of leaf

shallow serration around leaf margin

stout, angled twig, covered in downy hairs

FEATURES

Bark Gray-brown, smooth; becomes fissured in maturity

Fruit Shed large amounts of cotton-wool-like seeds

Populus lasiocarpa (Salicaceae)

CHINESE NECKLACE POPLAR

This broad-spreading tree has among the largest leaves of any poplar. It tends to be open and gangly, with long-spreading, widely spaced branches, and can appear ungainly in winter, when bare.

In fall, the leaves turn yellow, then crisp brown, making a clattering sound on windy days. In spring, stiff, yellow-green male and female flowers emerge in catkins borne on separate trees. The name "necklace" refers to the fruit, which appear as long, pendulous, green seed capsules in midsummer. These fruit ripen in late summer and release seeds to be dispersed by the wind.

♦ **Black cottonwood**
This taller, fast-growing tree (*Populus trichocarpa*) has smaller leaves that smell of balsam.

PROFILE

Lobed leaves
White poplar leaves vary in shape.
Some are ovate and irregularly
toothed, borne on short shoots.

- Up to 85 ft (26 m)
- Up to 70 ft (20 m)
- Deciduous
- Alternate simple
- Europe (except British Isles), W. Asia, N. Africa

leaf up to 4 in
(10 cm) long, 3 in
(8 cm) wide

shiny, dark
gray-green
upper surface

FEATURES

Bark Light gray, pitted with
small, black, diamond-shaped
patches when young

dense cover of
white hairs on
underside of leaf

Leaves Variously shaped;
can be maple-like, borne
on vigorous shoots

Populus alba (Salicaceae)

WHITE POPLAR

This large, broadly conical tree is often cultivated
in coastal locations because of its ability to cope with
onshore winds and salt spray. It is also planted alongside
freeways and highways, where it withstands atmospheric
pollution quite well.

The bark is coarsely cracked
with vertical fissures in maturity.
The leaves, borne alternately,
are attached to the shoots by a
1½ in- (4 cm-) long hairy petiole
(leaf stalk). Male and female
flowers are borne on separate
catkins before the leaves open.
Male catkins are crimson and gray
while the female are pale green,
both up to 3 in (8 cm) long. The
fruit is a green capsule, which
opens to release seeds covered
in fluffy, white hairs.

♧ Gray poplar
This hybrid (*Populus x
canescens*) has leaves with
a greenish white, feltlike
cover on the underside.

Crack willow leaves
Dark glossy green leaves are alternately arranged along stems. The leaves taper to a fine point.

↕ Up to 80 ft (25 m)

↔ Up to 70 ft (20 m)

✷ Deciduous

🍃 Alternate simple

◉ Europe, Asia

lanceolate leaf, up to 6 in (15 cm) long, 1 in (2.5 cm) wide

leaf tapers to fine point

FEATURES

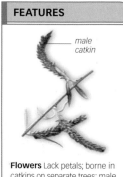

male catkin

Flowers Lack petals; borne in catkins on separate trees; male are yellow, females green

Salix fragilis (Salicaceae)

CRACK WILLOW

This broadly spreading tree is one of the largest willows. Often found growing alongside rivers and streams, it takes its name from the distinctive crack heard when a twig is snapped off the tree. The wood is brittle, and large branches are often broken off by high winds. The branches quite often take root where they touch the ground, and eventually another tree is formed, genetically the same as the parent.

The bark is dull gray, becoming heavily fissured and cracked from a relatively early age. Some stems may be as much as 6 ft (2 m) in diameter. The shoots are pale orange and conspicuous in late winter before new leaves appear. The leaves are lanceolate, tapering to a fine point. They are glossy dark green on the upper surface, with a raised central midrib, and blue-green on the underside. Small male and female flowers are borne in catkins.

♤ **Grey sallow (*Salix cinerea*)** This is a much smaller tree than the crack willow. It can also be distinguished from the latter by its hairy, obovate leaves and stout, hairy twigs.

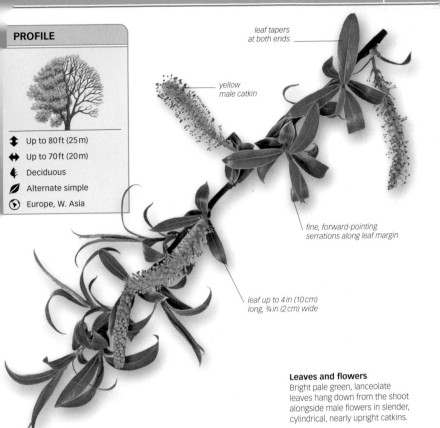

leaf tapers at both ends

yellow male catkin

fine, forward-pointing serrations along leaf margin

leaf up to 4 in (10 cm) long, ¾ in (2 cm) wide

PROFILE

↕ Up to 80 ft (25 m)

↔ Up to 70 ft (20 m)

🍂 Deciduous

🌿 Alternate simple

🌍 Europe, W. Asia

Leaves and flowers
Bright pale green, lanceolate leaves hang down from the shoot alongside male flowers in slender, cylindrical, nearly upright catkins.

FEATURES

Bark Brown-gray; develops coarse, vertical fissures with age

Shoots Slender, light gray-pink to olive-brown

Salix alba (Salicaceae)

WHITE WILLOW

A broadly columnar tree, the white willow is native across Europe and western Asia. This medium-sized tree thrives in damp soils, and is normally found growing alongside rivers and near water meadows.

A dense coating of white hairs on the leaves gives this tree a silvery appearance, particularly when the leaves are stirred by the wind. Young leaves are coated with silver hairs on both sides; mature leaves retain hairs on the underside. Both male and female flowers are clustered in catkins on separate trees. They grow up to 2½ in (6 cm) long in spring. Green female flowers soon become fluffy with white-haired seeds.

△ **Bay willow**
This tree (*Salix pentrandr*
has olive-green shoots
glossy green leaves, w
are aromatic when cr

PROFILE

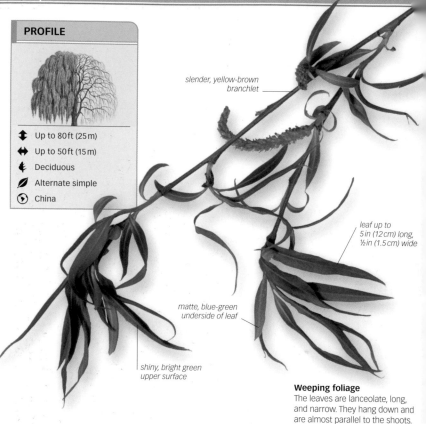

- Up to 80 ft (25 m)
- Up to 50 ft (15 m)
- Deciduous
- Alternate simple
- China

slender, yellow-brown
branchlet

leaf up to
5 in (12 cm) long,
½ in (1.5 cm) wide

matte, blue-green
underside of leaf

shiny, bright green
upper surface

Weeping foliage
The leaves are lanceolate, long,
and narrow. They hang down and
are almost parallel to the shoots.

FEATURES

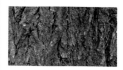

Bark Cracked and vertically
fissured, particularly in old age

flowers have
yellow anthers

Male flowers
Yellow-green, appear
with young leaves in
spring in slender catkins

Salix babylonica (Salicaceae)

WEEPING WILLOW

One of the best-known trees of the temperate world,
the weeping willow is commonly planted alongside rivers
and lakes. It was introduced into the West toward the
end of the 17th century and has long been cultivated
elsewhere. It is a wide-spreading, medium-sized tree
with a pendulous habit, featuring a wide head of
relatively level branches from which long, pendulous
branchlets descend almost vertically.

The leaves have shallow serrations along the margins.
Male and female flowers are borne in separate trees. In
recent years, this tree has been superseded by *Salix x
sepulcralis* 'Chrysocoma' (p.203), which is a cultivar of a
hybrid between the weeping willow (p.200) and the white
willow (p.199). The hybrid has brighter yellow twigs and
shoots, but is prone to attack from a canker disease,
which produces an open wound.

◊ **Dragon's claw willow (*Salix babylonica* var. *pekinensis*
'Tortuosa'**) Unlike the weeping willow, this smaller columnar
tree, up to 30 ft (10 m) tall, has twisted branches and twigs.

PROFILE

- Up to 30 ft (10 m)
- Up to 30 ft (10 m)
- Deciduous
- Alternate simple
- Europe, W. Asia

Goat willow catkins
This tree bears male flowers in golden catkins. Soft, silvery, female catkins known as pussy willow are borne separately on female trees.

tiny golden male flowers in catkins

FEATURES

Bark Dull gray; smooth, with minor fissures in maturity

Leaves Vary from oval to obovate or lanceolate, up to 4 in (10 cm) long

Salix caprea (Salicaceae)

GOAT WILLOW

Also known as pussy willow, this small, broadly spreading tree sometimes resembles a large shrub of bushy habit. It seldom grows more than 30 ft (10 m) tall. The goat willow is native to the whole of Europe and western Asia and is found on all soil types in both upland and lowland regions. It produces a prolific amount of seed, which is dispersed by the wind and can easily colonize bare or cultivated land some distance from the tree.

The tree's young shoots are covered in fine gray down. The leaves have a dull green and wrinkled upper surface and a gray-green underside, with dense gray hairs. Male and female flowers are borne in catkins on separate trees in spring. The fruit is a gray-green capsule covered in soft down, which opens to release numerous tiny, airborne seeds.

♧ **Common osier (*Salix viminalis*)** This tree is easily distinguished from the goat willow by its more shrubby habit with long, upright stems and longer, narrower leaves.

PROFILE

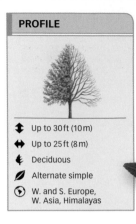

- Up to 30 ft (10 m)
- Up to 25 ft (8 m)
- Deciduous
- Alternate simple
- W. and S. Europe, W. Asia, Himalayas

light green midrib on leaf

leaf petiole up to ¼ in (0.5 cm) long

leaf up to 3 in (8 cm) long, ¾ in (2 cm) wide

Violet willow leaves
The leaves are shiny dark green on the upper surface and glaucous blue-green on the underside, with a shallowly serrated margin.

FEATURES

Bark Dark, almost purple-black; relatively smooth on older stems and trunks

Male flowers Borne in catkins, up to 2 in (5 cm) long

Salix daphnoides (Salicaceae)

VIOLET WILLOW

This broadly spreading tree is native to western and southern Europe, and western Asia, including the Himalayan region. It has long been cultivated for its plum- or violet-colored shoots which, when young, are covered with attractive white bloom. In cultivation, these trees are often coppiced every two to three years, so that multiple violet stems develop, growing up to 6–10 ft (2–3 m) tall. This is particularly effective in winter, when the stems are not masked by the foliage.

The lanceolate leaves remain green on the tree well into late fall before being shed. The leaf petiole is flushed crimson. White male catkins appear in early spring before the leaves emerge, becoming bright yellow as they mature. Gray-green female flowers are shorter and slender, ripening to release seed-bearing white "fluff" in late spring.

◊ **Magnificent willow (*Salix magnifica*)** Unlike the violet willow, this tree has oval or obovate magnolia-shaped leaves. It also bears attractive erect red and yellow catkins.

Leaves and flowers
The golden weeping willow bears leaves on long, pendulous, golden yellow secondary branches. The flowers, borne in catkins, appear along with leaves in mid-spring.

leaf ends in tapered point

golden yellow shoot

bright yellow catkin

PROFILE

↕ 70 ft (20 m)

↔ Up to 70 ft (20 m)

🍂 Deciduous

🌿 Alternate simple

📍 W. Asia

FEATURES

Leaves Lanceolate, slender; up to 5 in (13 cm) long, 1 in (2.5 cm) wide

Salix x sepulcralis 'Chrysocoma' (Salicaceae)

GOLDEN WEEPING WILLOW

A hybrid between the white willow (p.199) and the weeping willow (p.200), this tree developed naturally where the ranges of the two species met in western Asia. Its cultivated form 'Chrysocoma' has been propagated for its vibrant golden shoots and graceful pendulous habit and is often seen growing alongside rivers and lakes. This broadly spreading tree has ascending primary branches, which support drooping secondary branches reaching the ground.

The bark is pale gray-brown with shallow corky fissures. Shoots and young twigs are golden yellow, and are best seen in winter when low sunlight hits the crown. Both male and female flowers are usually borne in the same catkin, although occasionally the catkins are made up of all male flowers.

🔍 *Salix fargesii* This tree is distinguished by its small stature, red stems and leaf buds, and shiny mahogany-colored bark.

palmate leaf, up
to 8 in (20 cm) long
10 in (25 cm) wide

large,
toothed lobe

green fruit,
ripens to brown

stalk bears up
to four fruit

yellow veins usually
meet near leaf base

PROFILE

↕	100–130 ft (30–40 m)
↔	Up to 100 ft (30 m)
🍂	Deciduous
🌿	Alternate simple
🌐	Europe

Lobed leaves
The 3–5 lobed leaves of the
London plane are glossy,
dark green, and marked by
conspicuous yellow veins.

FEATURES

Bark Light brown; flakes in
large patches with age,
exposing creamy patches

tiny red
female
flower

Flowers Red female and
yellow male flowers borne
on same tree

Platanus x *acerifolia* (Platanaceae)

LONDON PLANE

Among the most common trees in towns and cities
across the temperate world, the sprawling London plane
is sometimes considered to be a form of the oriental
plane (p.205). The general consensus is that it is an
artificially created hybrid between the oriental plane and
the American buttonwood (p.206) that emerged sometime
before 1663. It can withstand
both atmospheric pollution and
pruning, and is thus commonly
found in many cities.

Its bark is light brown, flaking in
irregular scales. The leaves have a
bright green upper surface but
are paler on the underside, with
conspicuous yellow veins and
rusty hairs. The fruit are mace-
like, bristly spheres that remain
attached to the tree well after
the leaves have fallen.

Majestic tree
The London plane is a large,
broadly columnar, hybrid
tree of garden origin, never
found growing in the wild.

leaf up to 8in (20cm) long, 10in (25cm) wide

yellow-green stalk, up to 3in (7.5cm) long

globular, mace-like fruit, 1in (2.5cm) wide

Deeply lobed leaves
The glossy, deep green leaves of the oriental plane are deeply cut into five slender lobes. Globular fruit are borne on pendulous stalks.

FEATURES

Bark Buff-gray; flakes to reveal cream-pink patches

Platanus orientalis (Platanaceae)

ORIENTAL PLANE

Native to the eastern Mediterranean, the oriental plane is one of the largest temperate deciduous trees, with a trunk that can exceed 20ft (6m) in girth. It is also a long-lived tree—a tree on the Greek island of Kos is said to be of the time of Hippocrates, "the father of medicine."

The palmate leaves are shiny, rich green on the upper surface and pale green on the underside, with brown hairs along pronounced veins. In fall, they turn bright yellow and then dull gold before falling. Once on the ground, they take a long time to decompose. In late spring, the oriental plane bears tiny flowers in pendulous, rounded clusters. They are followed by fruit attached in clusters of two to six on pendulous stalks. These remain on the tree well into winter.

Large canopy
The oriental plane has a wide-spreading canopy made up of long and large-limbed branches.

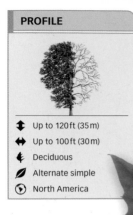

Up to 120 ft (35 m)

Up to 100 ft (30 m)

Deciduous

Alternate simple

North America

leaf up to 8 in (20 cm) long and wide

toothed leaf margin

glossy green upper surface of leaf

Palmately lobed leaves
The leaves have three angular, pointed lobes. They have a sharply toothed margin with hair-covered veins on the underside.

FEATURES

Bark Pale gray-brown trunk, smooth; flaking to reveal fresh, cream-colored bark beneath

Platanus occidentalis (Platanaceae)

BUTTONWOOD

Also known as American sycamore, this broadly spreading, large tree is native to North America, where it grows into a magnificent specimen with a broad, open crown, and a trunk up to 15 ft (5 m) in diameter. It is favored for planting in cities and parks, and provides shade in the summer months. When cultivated outside of North America, it seldom obtains the same noble stature, and often does not become well established. As with all members of the *Platanus* genus, it is tolerant of atmospheric pollution.

Unlike the bark on the trunk, the branches are blue-white, thin, and smooth. The leaves are bright, glossy green on the upper surface and pale green on the underside. New leaves are coated in soft, white down. Yellow male flowers and reddish female flowers appear in separate, rounded clusters on the same tree in late spring. The mace-like fruit are up to 1 in (2.5 cm) wide. They hang singly on thin stalks. Green when young, the fruit turn pink, and then brown when ripe.

PROFILE

tapering
leaf lobe

Leaves and fruit
Sweetgum leaves are palmately lobed and
slightly toothed around the margin. The fruit
is pendulous, long-stalked, and bristly.

leaf up to 6 in
(15 cm) long and wide

- ↕ Up to 150 ft (45 m)
- ↔ Up to 50 ft (15 m)
- ⚡ Deciduous
- ⬮ Alternate simple
- ⦿ E. US, C. America

FEATURES

Bark Dark brown, smooth;
becomes fissured into long
vertical ridges with age

Leaves Change color from
orange, through bright red to
purple in fall

Liquidambar styraciflua (Hamamelidaceae)

SWEETGUM

Also known as satinwood, this tree is native to the
East Coast and also parts of Central America. The
sweetgum tree is one of the main constituents of the
deciduous hardwood forests of North America. Its leaf
shape can be confused with maple, a member of
the *Acer* genus; however, maple
leaves are borne in opposite
pairs on a smooth shoot, whereas
the leaves of the sweetgum
are alternately positioned
on corky shoots.

The leaves display five to seven
tapering lobes, where the central
lobe is normally the largest. Small,
round, green-yellow male and
female flowers appear in late
spring. The fruit is up to 1 ½ in
(4 cm) wide, borne in clusters,
green ripening to brown.

Changing habit
The sweetgum has a
broadly conical shape
when young, but develops
a rounded habit with age.

deeply cut
oblong lobe

Lobed leaves
The palmate leaves are cut into
five three-lobed sections. They
are deep matte green on the upper
surface, with a paler underside.

leaf up to 3 in (8 cm
long and wide

slender petiole
(leaf stalk)

PROFILE

⬍ Up to 80 ft (25 m)

⬌ Up to 15 ft (5 m)

🍂 Deciduous

🌿 Alternate simple

◉ S.W. Turkey

FEATURES

Bark Orange-brown, thick;
cracks with age into small,
irregular-shaped plates

Leaves Turn from orange to
red to gold in fall

Liquidambar orientalis (Hamamelidaceae)

ORIENTAL SWEETGUM

This tree is commonly thought of as the Asian
version of the American sweetgum (p.207). However,
oriental sweetgum is native to southwest Turkey and
therefore not strictly Asian at all. In the wild, it is a
medium-sized, broadly conical tree. In cultivation,
it rarely attains half this size and is more bush-like
in its habit. However, what it lacks in stature, it makes
up for by providing some of the finest fall leaf colors
of any tree. The tree's resin has been used medicinally
for over 700 years.

The leaves have five deeply cut lobes, the central
lobe being the largest. Both male and female flowers are
yellow-green and very small. They are borne separately on
the same tree in spring. The rounded, pendulous fruit, up
to 1 in (2.5 cm) long, are similar to those of the sweetgum
(p.207). The fruit hang by a stalk up to 2 in (5 cm) long.

♀ **Chinese sweetgum (*Liquidambar formosana*)** This species
has duller, broader, green leaves with three to five lobes. They are
more angular and hairier on the underside and the petiole (leaf stalk).

PROFILE

- ↕ Up to 80 ft (25 m)
- ↔ Up to 50 ft (15 m)
- ❦ Deciduous
- ∅ Alternate simple
- ⊙ E. and S. North America

smooth leaf margin

tiny green flower

leaf up to 6 in (15 cm) long, 3 in (7.5 cm) wide

obovate to elliptic leaf with blunt tip

Lustrous leaves
The glossy leaves of the black gum are bright green on the upper surface and matte blue-green on the underside. Tiny flowers lie hidden amidst the leaves.

FEATURES

Bark Smooth, dark gray; cracks and fissures into rough square plates in maturity

orange-colored fall leaf

Leaves Turn yellow-orange to burgundy red in fall

Nyssa sylvatica (Cornaceae)

BLACK GUM

Also known as sour gum or black tupelo, black gum is native to eastern and southern North America from Ontario, Canada, in the north to New Mexico in the south. Since its discovery by Europeans in 1750, it has been extensively cultivated elsewhere and is often found in parks, gardens, and arboreta.

Both male and female flowers are small, green, and hard to see against the leaves. They are borne in long-stalked clusters on the same tree in summer. The edible fruit is a blueberry-colored, egg-shaped, glossy berry, up to ½ in (1.5 cm) long. Although edible, the fruit is not particularly palatable, being rather sour, hence its alternative name—sour gum.

Slender branches
The black gum has slender, spreading branches and a broadly conical to columnar crown.

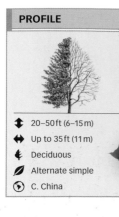

PROFILE

↕ 20–50 ft (6–15 m)
↔ Up to 35 ft (11 m)
❧ Deciduous
🖋 Alternate simple
⊙ C. China

leaf runs to a fine point

leaf up to 8 in (20 cm) long, 2½ in (6 cm) wide

Summer foliage
The leaves are shiny green on the upper surface and matte, pale green on the underside.

oblong-lanceolate leaf, entire around the margin

FEATURES

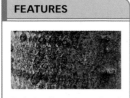

Bark Gray-brown, smooth; cracks and flakes with age

Nyssa sinensis (Cornaceae)

CHINESE TUPELO

This small, broadly conical tree is related to both the black gum (p.209) and the water tupelo (*Nyssa aquatica*). It is indigenous to central China, where it grows in woodland in the shade of other trees, and was introduced into western cultivation in 1902.

The leaves emerge green-red from winter buds, but the red quickly fades to deep green. They turn red, orange, and yellow in fall. The underside of the leaves have hairs on the margins and midrib. Male and female flowers are borne on the same tree in summer, in separate clusters, around leaf axils. They are five-petaled, small, and green, which makes them difficult to see among the foliage. The fruit is a purple-blue, egg-shaped berry up to 1 in (2.5 cm) long.

Showy fall foliage
The Chinese tupelo is a beautiful sight in fall, when the leaves turn a striking mix of colors.

shallowly toothed leaf margin

Bright green leaves
The ovate, at times elliptic, leaves are glossy bright green on the upper surface, and dull green, with hairs, on the underside.

leaf 5 in (12 cm) long, 2 1/2 in (6 cm) wide

FEATURES

colorful fall foliage

Leaves Turn yellow, orange, red, and purple in fall

Flowers Small clusters of ruby-red anthers; appear on bare branches in winter

Parrotia persica (Hamamalidaceae)

PERSIAN IRONWOOD

Persian ironwood is a popular tree in cultivation because of its beautiful fall leaf color. While in the wild it grows more upright, reaching 70 ft (20 m) tall, in cultivation, it is broadly spreading, occasionally reaching 50 ft (15 m) in height, and almost as much in width. It is named after German scientist and climber F. W. Parrot, who made the first European ascent of Mount Ararat in Turkey in 1829. In cultivation, the branches of this tree form an interlacing network reaching down to the ground, making it difficult to walk beneath the canopy.

The bark is dark brown and smooth. It flakes in patches to reveal light fawn-colored new bark. Leaves often display bronze-red margins when young. They have wavy margins near the tips. Flowers are small and petalless with red anthers. The fruit is a small, nutlike, brown capsule, up to 1/2 in (1 cm) wide.

♤ **X *Sycoparrotia semidecidua*** This hybrid between *Syco* *Parrotia* has leaves that are ovate, lighter green, and thinner *Sycopsis*. Its flowers are orange-red, with yellow anthers.

PROFILE

↕ 60–100 ft (18–30 m)

↔ Up to 70 ft (20 m)

♣ Deciduous

✿ Opposite simple or alternate simple

☉ Japan, W. China

Cordate leaves
The soft, thin leaves are heart-shaped, finely toothed around the margin, and as wide as they are long.

shallowly toothed margin

seed pod

petiole up to 2.5 cm (1 in) long

leaf up to 3 in (7 cm) long and wide

FEATURES

Bark Gray-brown, smooth; becomes shallowly fissured with vertical cracks in maturity

Leaves Orange to apricot-flushed red or sulfur yellow in fall

Cercidiphyllum japonicum (Cercidiphyllaceae)

KATSURA TREE

Katsura is one of the most attractive deciduous trees in the northern hemisphere. It is one of the two species in the family Cercidiphyllaceae, the other being *Cercidiphyllum magnificum*. After katsura's introduction into western cultivation in 1881, it became a popular tree for planting in parks, gardens, and arboreta.

The bark is freckled with lighter colored lenticels (pores). Held on lax, green-red petioles, the leaves are bright matte green on the upper surface and blue-green on the underside. However, by summer their upper surface also turns blue-green. In fall, they take on a variety of bright shades before eventually falling, giving off a sweet, caramel-like aroma as they begin to decompose. Bright red male flowers appear on side shoots before the leaves emerge in spring. Red female flowers develop on separate trees at the same time.

♤ ***Cercidiphyllum magnificum*** This is a smaller tree than the katsura tree, rarely reaching more than 30 ft (10 m) in height, with a smooth bark and larger, more coarsely toothed leaves.

PROFILE

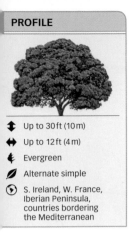

↕ Up to 30 ft (10 m)

↔ Up to 12 ft (4 m)

❅ Evergreen

⬭ Alternate simple

◉ S. Ireland, W. France, Iberian Peninsula, countries bordering the Mediterranean

glossy, dark green leaf, up to 4 in (10 cm) long

tough, leathery leaf

pale underside of leaf

fruit 1 in (2.5 cm) wide

uniform serrations around leaf margin

Ornamental species
With fruit ripening from flowers of the previous year, the Killarney strawberry tree bears fruit and flowers at the same time.

FEATURES

flower ½–¾ in (1.5–2 cm) long

Flowers Creamy white, urn-shaped, borne in drooping clusters

unripe fruit

Fruit Strawberry-like berry, green ripening to red

Arbutus unedo (Ericaceae)

KILLARNEY STRAWBERRY TREE

This broadly spreading tree has become popular for garden planting. The name "Killarney" refers to a part of Ireland, where this tree is extremely common. It is often used for hedging around fields in this region.

Young bark is a rich, red-brown color, which fades to a deep brown, and turns rough and fissured, giving the tree a gnarled appearance. The leaves are elliptic to obovate in shape. The flower clusters are up to 2 in (5 cm) long and are borne at the ends of shoots in fall. The warty fruit ripens from green through yellow to red in fall from flowers produced in the previous year. Though the fruit is edible, it tastes dry and is not particularly enjoyable.

Garden tree
The Killarney strawberry tree is small, with a broadly spreading habit and glossy green foliage.

elliptic to oval leaf with smooth margin

flower up to ¼ in (6 mm) long

leaf up to 4 in (10 cm) long, 2 in (5 cm) wide

Pitcher-like flowers
Panicles of small flowers appear on short stalks in spring. Green-tinged at first, they open white at the ends of shoots, in upright clusters.

FEATURES

Bark Red-brown to cinnamon; peels in ragged strips to reveal orange or cream new bark

Arbutus andrachne (Ericaceae)

GRECIAN STRAWBERRY TREE

Also known as Cyprus strawberry tree, this small, broadly spreading tree is indigenous to southeast Europe, and has been in cultivation in warm, temperate regions of the world since 1724. Its canopy is dome-shaped, and it often develops low branching, and multiple stems—sometimes from just above ground level, giving it a shrublike appearance. When it is grown alongside Killarney strawberry tree (p.213), the two may cross-pollinate to produce the hybrid strawberry tree (*Arbutus* x *andrachnoides*).

The leaves are thick and leathery, glossy dark green on the upper surface, pale green on the underside, and with no serrations around the leaf margin. The fruit is a rounded red-orange berry, up to ½ in (1 cm) wide.

◊ **Hybrid strawberry tree (*Arbutus* x *andrachnoides*)** Hardier than the Grecian strawberry tree, this tree has brighter, cinnamon-red bark and shoots, and it flowers in late fall rather than spring.

PROFILE

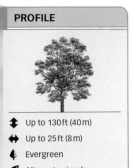

↕ Up to 130 ft (40 m)

↔ Up to 25 ft (8 m)

🌿 Evergreen

⬤ Alternate simple

◉ West coast of North America

upright, urn-shaped flower

thick, elliptic, leathery leaf

Large flower clusters
Tiny, creamy white flowers are borne in erect panicles at the ends of shoots in spring.

dark green leaf, up to 6 in (15 cm) long, 3 in (7.5 cm) wide

FEATURES

Bark Peels in papery flakes to reveal fresh green or cream color beneath

fruit up to ½ in (1.5 cm) wide

Fruit Rounded, strawberry-like; changing from green to orange to red in early fall

Arbutus menziesii (Ericaceae)

MADRONE

This broadly columnar tree is called the Pacific or Californian madrone. It is the largest member of the Ericaceae family, growing up to 130 ft (40 m) in the wild, though it normally remains around half this height in cultivation. It is indigenous to the west coast of North America, from British Columbia, including Vancouver Island, Canada, south to California. Madrone was introduced into Europe by Scottish plant collector David Douglas in 1827.

It has smooth and attractive red-pink bark, sometimes purple toward the base, which peels in papery flakes. The glossy leaves have pale blue-white undersides, and are borne on red petioles. The rough, strawberry-like fruits are covered in warts.

Elegant evergreen
The madrone is widely planted in parks, gardens, and arboreta across the northern temperate world.

Flowers and leaves
The fragrant flowers hang down in clusters, or singly, along the underside of the branches. Leaves are rich green, oval to elliptic.

small, shallow teeth on leaf margin

golden anthers

bell-shaped, creamy white flower

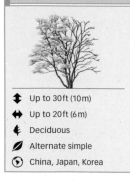
Styrax japonicus (Styracaceae)

JAPANESE SNOWBELL

As the name suggests, this tree is native to Japan. However, Japanese snowbell is endemic to parts of China and Korea as well. It was introduced in the West in 1862, when botanical collector Richard Oldham brought a specimen from Japan. This was subsequently planted in the Royal Botanic Gardens at Kew, London.

The orange-brown bark develops fissures when the tree is around 30 years old. The leaves are 4 in (10 cm) long and 2 in (5 cm) wide, with a slight sheen on the upper surface. They are pale green and matte on the underside, and normally arranged in groups of three. In early summer, the canopy is covered in open flowers. Once pollinated, these develop into egg-shaped, gray-green fruit, each containing a single seed.

Spreading form
The Japanese snowbell often has a short trunk and widely spreading branches that emerge low on the trunk.

small, slender flower, up to 1 in (2.5 cm) long

bright golden anthers

roughly oval leaf with rich green upper surface

leaf up to 5 in (13 cm) long, 4 in (10 cm) wide

Spreading raceme
Chinese snowbell tree bears white flowers, borne on terminal, upright, long racemes, which become lax and spreading over time.

PROFILE

↕ Up to 25 ft (8 m)

↔ Up to 10 ft (3 m)

🍂 Deciduous

🌿 Alternate simple

☉ C. and W. China

Styrax hemsleyanus (Styracaceae)

CHINESE SNOWBELL TREE

This slender, open-branched tree is also known as Hemsley's storax, after William Hemsley, one-time keeper of the herbarium at the Royal Botanic Gardens, Kew, London. Native to central and western China, this tree grows on mountain slopes at an altitude of 8,850 ft (2,700 m) above sea level. It was introduced into western cultivation by English plant collector Ernest H. Wilson in 1900.

The Chinese snowbell tree has a dark brown and relatively smooth bark, even in maturity. The leaves are ovate to obovate, with pale green undersides, sometimes covered in hairs. The leaves have occasional blunt-edged serrations around the margins. A distinguishing feature of this species is its exposed, bright orange to chocolate-brown leaf bud. White flowers, occasionally with a slight pink blush, emerge from the ends of short shoots in late spring to early summer. Flowers are followed by down-covered, ovoid, gray-green fruit.

flower up
to 1¼in
(3cm) long

leaf up to
8in (20cm) long

yellow anthers
on white flower

Broadly ovate leaves
On young trees, the large leaves
can be irregularly shaped, but they
are broadly ovate to orbicular on
mature trees.

PROFILE

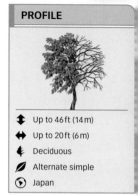

⬍ Up to 46ft (14m)

⬌ Up to 20ft (6m)

🍂 Deciduous

🌿 Alternate simple

◉ Japan

FEATURES

Fruit Gray-green, egg-shaped
berry, carrying a single seed

Styrax obassia (Styracaceae)

FRAGRANT SNOWBELL

Also known as bigleaf storax, this tree is indigenous
to Japan. It was introduced into cultivation by British
botanist Charles Maries in 1879. Today, despite its
hardiness and beauty, it is not frequently cultivated.

The gray bark turns brown and
fissured in maturity. In the second
year, the bark on its shoots is
chestnut-brown and flaky. Bright
green on the upper surface,
leaves have a paler underside
with white down. Leaf petioles
swell at the base to enclose the
leaf bud. Fragrant, bell-shaped
flowers are borne in lax, green,
terminal racemes up to 8in
(20cm) long, in early summer.
These are followed by ovoid fruit,
up to ½in (1.5cm) long and
covered in pale brown hairs.

Arching canopy
The fragrant snowbell's
open branches arch at
the canopy to give it a
round-headed appearance.

PROFILE

- Up to 70 ft (20 m)
- Up to 30 ft (10 m)
- Deciduous
- Alternate simple
- S.E. United States

Bell-shaped flowers
The white flowers hang in groups of three or four, along the underside of the branches in mid- to late spring.

flower borne on long stalk

bright orange stamens

white, bell-shaped flower

FEATURES

leaf up to 8 in (20 cm) long, 4 in (10 cm) wide

Leaves Ovate to oblong; finely toothed along the margin

slender beak tip

Fruit Pendulous, angular capsule, attached to tree by ¾ in- (2 cm-) long stalk

Halesia carolina (Styracaceae)

CAROLINA SILVERBELL

Also known as Carolina snowdrop tree, this small, broadly conical tree is native to the southeast United States, from Virginia south to Florida. It was discovered and introduced into cultivation in 1756. Since then, it has been widely planted elsewhere in the US and in other parts of the western world. The botanical name *Halesia* commemorates the English clergyman and botanical author Dr. Stephen Hales.

The bark of the tree is pale brown and smooth, becoming darker and scaly in maturity. The ovate to oblong leaves taper to a long point. They are bright green on the upper surface and pale green on the underside, with some hairs, especially around the leaf veins. In fall, the leaves turn golden yellow before being shed. Bell-shaped flowers are followed by pear-shaped, four-winged fruit capsules, which are green at first, ripening to brown.

PROFILE

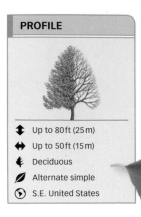

- Up to 80 ft (25 m)
- Up to 50 ft (15 m)
- Deciduous
- Alternate simple
- S.E. United States

leaf has long, pointed tip

leaf up to 10 in (25 cm) long, 4 in (10 cm) wide

finely serrated leaf margin

deeply impressed veins

Leaves and fruit
The ovate to oblong leaves are borne alternately on the shoot. The pendulous, capsule-like fruit is up to 1½ in (4 cm) long and borne on a ¾ in- (2 cm-) long stalk. It ripens from green to brown.

four-winged, green capsule-like fruit

FEATURES

Bark Pale gray, with shallow, widely spaced, pink fissures; turns darker gray in maturity

Flowers Crocus-shaped; hang in clusters, normally of three

Halesia monticola (Styracaceae)

MOUNTAIN SNOWDROP TREE

This broadly spreading tree is native to the mountains of the southeast United States, from North Carolina to Georgia. It has been widely cultivated elsewhere, in the US and around the world, since its introduction into European cultivation in about 1897. It is distinguished from the Carolina snowdrop tree (p.219) by its larger size, flowers, and fruit.

The shoot is gray-brown, and the leaf bud small and deep purple. The leaves are bright green on both sides, and turn yellow in fall. White flowers, 1¼ in (3 cm) long, emerge from pale pink buds, borne on ¾ in- (2 cm-) long stalks. They hang beneath the branches in profusion, appearing in late spring and early summer.

Cultivated habit
The mountain snowdrop tree grows as a large shrub or a small-spreading tree in cultivation.

PROFILE

- 30–50 ft (10–15 m)
- Up to 25 ft (8 m)
- Deciduous
- Alternate simple
- Japan

Camellia-like flowers
These white, frilly petaled, open flowers appear singly in leaf axils in summer. When old, they fall in one piece from the tree.

pale, shiny leaf underside

bright orange-yellow anthers

matte, dark green upper surface of leaf

roughly oval leaf, up to 4 in (10 cm) long, 1½ in (4 cm) wide

FEATURES

Bark Smooth, red-brown; flakes to reveal cream, pale orange, or light brown patches

Stewartia pseudocamellia (Theaceae)

JAPANESE STEWARTIA

A member of the same family as the *Camellia*, this small to medium-sized tree produces very similar flowers. It is native to Japan and was introduced into western cultivation in the 1870s. It takes its generic name from John Stuart, the Earl of Bute and a patron of the Royal Botanic Gardens in Kew, London. For this reason, some books refer to it as Stuartia.

The tree becomes broadly columnar with age. The leaves are elliptic to ovate, finely toothed, and borne alternately on twigs. They sometimes have hairs in the vein axils on the underside. In fall, the leaves range in color from orange-red to purple before falling. The Japanese stewartia flowers in mid- to late summer, and the flowers are shed from the tree in one piece, rather than falling petal by petal. Its fruit is a woody, triangular, red-brown capsule, up to ¾ in (2 cm) long.

⚲ **Sawtooth pseudocamellia (*Stewartia serrata*)** This tree is distinguished by the neat, forward-pointing serrations on its leaf margins, and smaller flowers with a red blotch at the base.

PROFILE

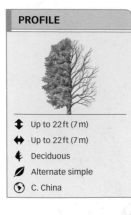

↕ Up to 22 ft (7 m)

↔ Up to 22 ft (7 m)

🌿 Deciduous

🍃 Alternate simple

◉ C. China

leaf up to 4 in (10 cm) long, 1½ in (4 cm) wide

pronounced vein on leaf

bright green leaf upper surface

bright orange anthers

Cup-shaped flowers
Creamy white flowers emerge in midsummer bearing frilly edged petals and clusters of bright orange anthers.

FEATURES

Bark Red-brown and smooth; flakes to reveal patches of fresh bark beneath

pale green, bulbous bud

Buds Green, short-stalked, carried solitary in leaf axils

Stewartia sinensis (Theaceae)

CHINESE STEWARTIA

This small deciduous tree, sometimes a large shrub, is native to Hubei province in central China. It was discovered and introduced into western cultivation by the English plant collector Ernest H. Wilson in 1901. Since then, it has been widely planted in gardens for its bark coloring, attractive summer flowers, and vibrant fall leaf tints.

The thin bark flakes and peels to reveal patches of cream, fawn, or light gray fresh bark beneath. Densely borne leaves are bright green with a sheen on both surfaces. Elliptic to elliptic-oblong in shape, they display distinctly pronounced veins and a long, pointed, and curved tip. The leaves are alternately attached to crimson-purple shoots by a ½ in- (1 cm-) long, red, and slightly hairy petiole (stalk). In fall, they turn bright crimson, orange, and yellow before falling. Upright flower buds open in midsummer to release beautiful camellia-like flowers up to 1½–2 in (4–5 cm) wide. The white flowers are slightly fragrant.

Honey flower
These ivory-white flowers, cup-shaped at first, open up to 1½ in (4 cm) wide, and contain highly prized, rich nectar.

wavy-edged petal

dark green leaf upper surface, with slight sheen

purple-tipped, orange stamens

pale leaf underside, with coating of fine, white hairs

PROFILE

- 30–70 ft (10–20 m)
- Up to 25 ft (8 m)
- Evergreen
- Opposite simple
- Chile, Argentina

Eucryphia cordifolia (Eucryphiaceae)

ULMO

This elegant tree grows in the foothills of the Andes mountain range, at an altitude of up to 2,300 ft (700 m) above sea level. It was first cultivated in Europe around 1851. In its native Chile and Argentina, it is currently threatened by logging and loss of habitat.

This densely branched tree is broadly conical. Its dark chocolate-brown, smooth bark is horizontally marked with fawn-colored lenticels (raised pores). The leaves are oblong, wavy-edged with shallow, blunt serrations, occasionally heart-shaped at the base, and run to a broad, rounded tip. They are up to 3 in (8 cm) long and 1½ in (3.5 cm) wide. Four-petaled flowers, borne singly in leaf axils, appear in late summer. Purple-tipped stamens give the center of the flowers a speckled appearance. The flowers' rich nectar is commercially harvested by beekeepers and sold as ulmo honey.

🌣 **Pinkwood (*Eucryphia moorei*)** With smaller flowers and pinnate leaves consisting of up to 13 slender leaflets, this Australasian tree is easily distinguished from the ulmo.

Striking flowers
Creamy or ivory-white flowers
are produced in late summer.
Sometimes these last well
into early fall.

*conspicuous
stamens
and styles*

*irregularly toothed
around leaf margin*

*flower up to
2½ in (6 cm) wide*

*elliptic leaflet, up to
2½ in (6 cm) long,
1¼ in (3 cm) wide*

PROFILE

↕ 20–30ft (6–10m)

↔ Up to 20ft (6m)

🌢 Evergreen/Deciduous
in cultivation

▱ Opposite compound

⊙ Chile

Eucryphia glutinosa (Eucryphiaceae)

EUCRYPHIA

This small, neat tree is evergreen in its native
Chile but usually deciduous in cultivation in Europe.
It has a broadly columnar form, with a mass of
strongly erect branches. Eucryphia, sometimes
known as brush bush, was introduced into western
cultivation in 1859, and is considered one of the most
beautiful flowering trees.

Its bark is dark brown and relatively smooth even in
old age. The leaves are pinnate, made up of three to five
leaflets. Each leaflet has a prominent wavy margin, a shiny
dark green upper surface, and a pale green underside. In
fall, the leaves turn brilliant orange-red before falling. The
four-petaled flowers emerge in the leaf axils in late
summer and are normally borne in profusion on the tree.
The fruit is an oval, woody capsule, up to ½ in (1.5 cm)
long, and contains several seeds.

🌢 **Nyman's eucryphia (*Eucryphia x nymansensis*)** This garden
cultivar has both simple and compound leaves on the same tree.
It is more upright and freer flowering than eucryphia.

elliptic or obovate leaflet, up to 4 in (10 cm) long

pealike flower, up to 1 in (2.5 cm) wide

dense, pendulous raceme, up to 12 in (30 cm) long

PROFILE

⭥ Up to 28 ft (9 m)

⭤ Up to 15 ft (5 m)

❦ Deciduous

🖉 Alternate compound

⊙ C. and S. Europe

Flowers and leaves
Golden yellow, pendulous flowers are borne in profusion in spring. The trifoliate leaves are green.

FEATURES

Fruit Green, hairy, bean-like seed pods, contain several small, round black seeds

Laburnum anagyroides (Fabaceae)

COMMON LABURNUM

Probably one of the most familiar trees in cultivation, the common laburnum is small and broad-spreading. Native to southern and central Europe from France to Hungary and Bulgaria, it naturally occurs in mountainous regions at altitudes of up to 6,500 ft (2,000 m). It is also short-lived—between 50 and 60 years is considered a good age.

The bark is brown-gray and smooth, becoming shallowly fissured in maturity. Each leaf is composed of three leaflets, rich green on the upper surface, gray-green on the underside and covered with silver hairs when young. They are attached to the shoots by a short petiole (stalk). New shoots are olive-green, while winter buds are pale gray-brown and covered in silvery hairs. After flowering, seed pods develop and ripen to brown. They crack open on the tree to reveal poisonous seeds.

♀ **Voss's laburnum (*Laburnum x watereri*)** This hybrid between *L. alpinum* and *L. anagyroides* produces larger flowers, borne in greater profusion. The leaves are slightly more hairy.

Drooping flowers
Flowering in early summer, this tree has several small, pealike flowers, widely spaced on slender racemes, which droop downward from slender shoots.

golden yellow flower, up to 1 in (2.5 cm) long

FEATURES

leaflet up to 5 in (13 cm) long

Leaves Trifoliate, deep shiny green on both surfaces

Laburnum alpinum (Fabaceae)

ALPINE LABURNUM

Also known as the Scotch laburnum, this small, widely spreading tree is native to the mountains of central and southern Europe and has been cultivated elsewhere since the end of the 14th century. It copes better with harsh winter conditions than the common laburnum (p.225), hence it is frequently planted along roads and tracks in Scotland—from where it gets its alternative name.

The bark is brown-gray and smooth, becoming only shallowly fissured in maturity. New shoots are olive-green bearing pale gray-brown winter buds covered in silver-gray hairs. Trifoliate leaves are attached to shoots by a short green petiole (stalk). Each leaflet is up to 4 in (10 cm) long and 2 in (5 cm) wide. Flowers are borne in hanging panicles up to 16 in (40 cm) long. The fruit are flat, green seed pods, which ripen to brown. They contain hard-coated, round, black, poisonous seeds.

♤ **Mount Etna broom (*Genista aetnensis*)** Slender, green, almost leafless shoots and yellow, pealike flowers, borne in erect clusters along shoots, distinguish this tree from the Alpine laburnum.

PROFILE

↕ Up to 30ft (10m)

↔ Up to 15ft (5m)

❦ Deciduous

▱ Alternate simple

◉ Middle East

leaf up to 5in (12cm)
long and wide

leaf upper surface
yellow-green to
blue-green

pale blue-green
leaf underside

bright, rosy pink
flowers, borne in
clusters of 3–6

Flowers on old wood
As rounded to heart-shaped
leaves unfurl in spring, clusters
of bright pea-flowers are borne
along old shoots.

FEATURES

Bark Gray-brown, creased;
fine brown fissures in maturity

green when
young, ripening
to brown

Fruit Long, thin, purple-tinted
seed pod

Cercis siliquastrum (Fabaceae)

JUDAS TREE

This small, broadly spreading deciduous tree is said
to be the tree on which Judas Iscariot hanged himself
after betraying Christ. However, its name is more likely
to have been derived from "Judea's tree," referring to
the region of its origin, now Israel and Palestine, where the
tree occurs commonly. It has long been cultivated across
Europe for its bright pink flowers,
and it grows particularly well in
hot, dry, sunny conditions. Older
trees may develop a pronounced
leaning habit.

The broad leaves are held on
pink-red shoots by a pink petiole,
from which light green-yellow
leaf veins radiate. In summer, the
flowers grow into conspicuous
seed pods, up to 5in (12cm) long
and ¾in (2cm) wide, which stay
on the tree until winter.

♤ **Eastern redbud**
This small, round-headed
tree (*Cercis canadensis*) has
paler flowers and thinner,
bright green leaves.

soft needlelike,
untoothed leaflet

dense flower
clusters with
salmon-pink
stamens

flowers borne
on erect stalk

Leaves and flowers
Bipinnate leaves are up to 20 in
(50 cm) long. Borne on soft, green
shoots, they give the tree a
"feathery" appearance. Fragrant
flowers appear in summer.

PROFILE

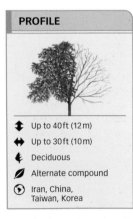

↕ Up to 40 ft (12 m)

↔ Up to 30 ft (10 m)

🍂 Deciduous

🌿 Alternate compound

📍 Iran, China,
Taiwan, Korea

Albizia julibrissin (Fabaceae)

SILK TREE

Sometimes mistaken for the silver wattle (p.229), this broadly spreading, flat-topped tree is native to Iran, China, Taiwan, and Korea. Also known as the mimosa or pink siris, it was introduced into western cultivation by the English botanist Ernest H. Wilson in 1918. The silk tree is a beautiful sight when in flower, making it a popular street plant in the United States. However, the tree is not completely hardy and struggles in the northern US and in parts of northern Europe.

The bark of the silk tree is dark brown and smooth. Foliage is similar to the silver wattle, with leaves made up of numerous small, tapering, bright green leaflets, each up to ½ in (1 cm) long. The flowers are borne on erect stalks in late summer to early fall. They have fluffy clusters of salmon-pink stamens that fade to lemon at the base, making them look like pink powder puffs from a distance. The fragrance of the flowers is like that of freshly cut hay. Seeds are borne in flat, light brown pealike pods, up to 6 in (15 cm) long.

PROFILE

- 70–80 ft (20–25 m)
- Up to 30 ft (10 m)
- Evergreen
- Alternate compound
- S.E. Australia, Tasmania

small, linear leaflet

Vibrant flower cluster
The tiny flowers have sulfur-yellow stamens, and are borne on a panicle up to 4 in (10 cm) long.

flower up to ½ in (1 cm) wide

bipinnate leaf

blue-green shoot

FEATURES

flattened seed pod

Fruit Glaucous green seed pod, ripening from green to brown

Acacia dealbata (Fabaceae)

SILVER WATTLE

Also known as mimosa, this tender, temperate, medium-sized tree is prized by florists the world over for its delicate, feathery foliage and fragrant, yellow flowers. Native to southeast Australia and Tasmania, this broadly conical tree is a popular ornamental street tree in Mediterranean countries, but does not grow as well in cooler climates farther north.

The bark is smooth, ranging between gray-green to almost glaucous blue when young, and turning darker and more brown in maturity. The leaves have a matte, mid-green upper surface and a sage-green underside. In the southern hemisphere, the tree blooms in summer, while in the north, this usually translates into early spring. The flowers are rounded with numerous yellow stamens, which are slightly fragrant. Its fruit is a seed pod containing several round, brown seeds.

♧ **Cootamundra wattle (*Acacia baileyana*)** Although similar to the silver wattle, this tree tends to be smaller and with many branches. Its smaller, bipinnate leaves are slightly more glaucous.

↕ Up to 130 ft (40 m)

↔ Up to 70 ft (20 m)

🌿 Evergreen

🍃 Transitions from compound to simple

🌐 E. Australia, Tasmania

undivided adult leaf

elliptic to oblanceolate mature leaf

fragrant globular flower, up to ½ in (1 cm) wide

Adult foliage
In maturity, blackwood acacia leaves are alternate and undivided. The flowers have rounded heads and contain numerous pale yellow stamens.

FEATURES

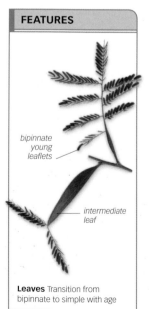

bipinnate young leaflets

intermediate leaf

Leaves Transition from bipinnate to simple with age

Acacia melanoxylon (Fabaceae)

BLACKWOOD ACACIA

Also known as the sally wattle or the Australian blackwood, this pyramid-shaped tree reaches up to 130 ft (40 m) in height, though it normally remains much shorter. It has a range that extends from southern Queensland to central Tasmania. It grows at 3,280 ft (1,000 m) above sea level in the Tasmanian rainforest.

The straight trunk has dark brown bark, smooth when young, vertically cracking into flaking plates with age. The tree is sometimes coppiced (cut back) and stumps may produce dull green phyllodes (flattened leaf stalks). Bluish white shoots bear leaves that are mid-green on the upper surface and blue-green on the underside. Bipinnate young leaves give the tree a feathery appearance. The mildly fragrant flowers are borne on panicles (branched flower stalks) up to 4 in (10 cm) long and are followed by reddish brown fruit pods that contain flat, rounded seeds.

♧ **Oven's wattle (*Acacia pravissima*)** Unlike blackwood acacia, this large shrub or small tree has arching branches and triangular phyllodes that are more prominent than the leaves.

PROFILE

Crinkly flowers
Deep lilac-pink flowers with
crinkly petals are borne in
branched flower clusters.

*flowers held in
panicle, up to
6 in (15 cm) long*

↕ Up to 20 ft (6 m)

↔ Up to 20 ft (6 m)

❧ Deciduous

🌿 Opposite or alternate
or borne in whorls
of three

◉ China, Korea,
N.E. India

FEATURES

*smooth leaf
margin*

Leaves Obovate; up to 2 in
(5 cm) long, 1 in (2.5 cm) wide

Lagerstroemia indica (Lythraceae)

CRAPE MYRTLE

This wide-spreading, flat-topped tree is native to
China, Korea, and parts of northeast India. It is widely
planted on streets and in parks and gardens throughout
the southern states of the US, India, and much of the
Mediterranean. Farther north, it is rather tender and
requires shelter and a sunny area.

The bark is thin, smooth, and light fawn. It flakes
to reveal patches of mottled
gray, pink, and cinnamon
beneath. The trunk can have
multiple stems, and both the
trunk and the main branches
are sometimes angular or
ribbed, rather like the common
hornbeam (p.184). The leaves
are deep green, thick, tough,
and privet-like. In fall,
they turn red or orange
before being shed.

Summer flowers
The broad canopy of crape
myrtle bears flowers in
profusion from midsummer
to early fall.

Wild cherry blossoms
Borne in clusters, the fragrant flowers open in spring, just before or as the leaves emerge.

elliptic leaf, up to 6 in (15 cm) long

large, white, five-petaled flower

flower up to 1¼ in (3 cm) wide

FEATURES

Bark Gray-red, smooth, turning red-brown when mature

reddish purple berry

Fruit Round, shiny cherries, up to ½ in (1.5 cm) wide

Prunus avium (Rosaceae)

GEAN

Also known as wild cherry or mazzard, this hardy, broadly columnar tree has a natural range spanning Europe and western Asia. It commonly grows on the edge of woodland or in small clearings, but is also likely to be found in city parks, highway embankments, and farmland copses. Gean is the parent of many domestic sweet cherries grown for their edible fruit.

In maturity, the bark turns glossy, sometimes peeling with light brown horizontal bands of lenticels (pores). The leaves, up to 2 in (5 cm) wide, have a sharply serrated margin and end in a short point. They are deep green on the upper surface and pale green on the underside. They turn orange, red, or yellow in fall. A bitter fruit is borne in summer.

⌂ St. Lucie cherry
This tree (*Prunus mahaleb*) can be told apart from gean by its darker bark, smaller stature, and rounded habit.

Serrated leaves
The bird cherry tree's dark matte green leaves are finely serrated around the margin and run to a short point at the tip.

leaf ends in abrupt point

leaf up to 4 in (10 cm) long

FEATURES

Bark Smooth, dark gray; emits an acrid scent when scratched

Flowers White, slightly fragrant, borne in upright racemes in mid- to late spring

Prunus padus (Rosaceae)

BIRD CHERRY

This deciduous, hardy, small to medium-sized tree is extremely beautiful when in bloom, and has given rise to a number of ornamental cultivars, which are now popularly grown in gardens across the northern hemisphere. As the name suggests, this tree produces a cherrylike fruit, which is relished by birds but is bitter to the human palate.

The leaves turn orange-red in fall before being shed. Small flowers are borne in upright racemes, 4–6 in (10–15 cm) long. These become lax, spreading, and pendulous as the flowers open. The flowers are followed by small, round to egg-shaped berries, up to ½ in (1 cm) wide. Initially green, the berries ripen to red, then black.

Spreading habit
With a conical habit in youth, the bird cherry becomes rounded, with drooping branches, in maturity.

PROFILE

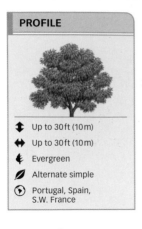

⬍ Up to 30 ft (10 m)

⬌ Up to 30 ft (10 m)

🍃 Evergreen

🌿 Alternate simple

⊙ Portugal, Spain, S.W. France

flowers borne in long, narrow raceme, up to 10 in (25 cm) long

creamy white flower, with five petals

ovate, elliptic leaf, tapers to short point

shallow, rounded serrations along leaf margin

Flowers and leaves
Creamy white flowers are reminiscent of the lily of the valley. Leaves are dark green on the upper surface and paler on the underside.

FEATURES

Fruit Egg-shaped; green, ripening through red to black

Prunus lusitanica (Rosaceae)

PORTUGUESE LAUREL

A small to medium-sized tree, the Portuguese laurel is perhaps taller in its natural habitat. It is grown around the Mediterranean region as a single specimen, like bull bay (see p.96), and is quite often clipped into architectural forms. Elsewhere, it is grown as a dense, luxuriant hedge or screen. It has been cultivated in this way since the 17th century, or perhaps even earlier.

The bark is dark slate-gray to gray-brown and relatively smooth even in maturity. Dark glossy green leaves are up to 4 in (10 cm) long and 2 in (5 cm) wide. Small flowers borne in early summer are scented, although not necessarily pleasantly. Long clusters of fruit follow the flowers. Individual fruits are widely spaced along the stalk.

♤ **Broadly spread habit**
The evergreen Portuguese laurel is shrubby and multi-stemmed, and serves well as a hedge or screen.

PROFILE

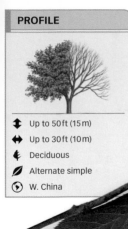

⬍ Up to 50 ft (15 m)

⬌ Up to 30 ft (10 m)

❧ Deciduous

🍃 Alternate simple

◉ W. China

willowlike leaf, up to 4 in (10 cm) long

egg-shaped, crimson berry, borne singly or in pairs on long stalks

finely toothed leaf running to long, pointed tip

Crimson cherry
The fruit of the paperbark cherry emerges as a yellow berry in fall, ripening to crimson-red.

FEATURES

Bark Deep-red, smooth; peels occasionally

flower up to ¾ in (2 cm) wide

orange-tipped stamen

Flowers White; produced in spring after leaves appear

Prunus serrula (Rosaceae)

PAPERBARK CHERRY

Native to Tibet and other parts of western China, this compact but broadly spreading, small, deciduous tree was first introduced into western cultivation by British plant collector Ernest H. Wilson in 1908. It reaches up to 50 ft (15 m) in the wild, but is usually shorter in cultivation.

The paperbark cherry is generally considered to be among the most attractive of all trees due to its bark, which looks like highly polished mahogany furniture. It is divided into horizontal sections by bands of light brown stomata (microscopic pores) and is particularly striking under the winter sun, when the tree is bare. Small, white, and relatively inconspicuous flowers appear in spring, while tiny, berrylike, purely ornamental fruit emerge in fall.

◇ **Himalayan cherry (*Prunus rufa*)** Its peeling, reddish brown, ornamental bark is similar to that of paperbark cherry. It has rusty haired young shoots and small clusters of pale pink flowers.

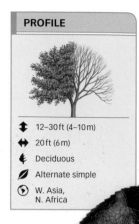

PROFILE

- 12–30 ft (4–10 m)
- 20 ft (6 m)
- Deciduous
- Alternate simple
- W. Asia, N. Africa

leaf up to 5 in (12 cm) long

oval to lanceolate leaf, with serrated margin

leaf held by 1 in (2.5 cm) petiole

fruit up to 2 in (5 cm) long

flesh splits open as fruit ripens

Leaves and fruit
A fleshy, yellow fruit covered with velvety hairs encloses the edible almond seed.

FEATURES

flower 1¼–2 in (3–5 cm) wide

Flowers Borne singly or in pairs in early spring

Seeds White, edible; carried in woody shells

Prunus dulcis (Rosaceae)

ALMOND

Known widely for its fruit, this small tree has a natural distribution extending across northern Africa and western Asia. However, since the Roman period it has been cultivated elsewhere and is now naturalized throughout the Mediterranean region and southern Europe.

The gray-black bark is smooth at first, cracking into small squares on maturity. Green, young twigs turn purple and then gray. Dark green to yellow-green leaves may appear curled and wrinkled due to a fungal disease known as peach leaf curl. Five-petaled flowers are white, sometimes faintly blushed pink, and produced singly or in pairs in early spring before the leaves emerge. The fruit is covered in a gray-green, thick, leathery coat called a hull. Inside the hull is a hard shell, or stone, which contains the edible seed.

♧ **Apricot (*Prunus armeniaca*)** This tree mainly differs from almond in its edible, fleshy, yellow fruit, which ripens to orange-red, and its ovate-shaped smaller leaves.

PROFILE

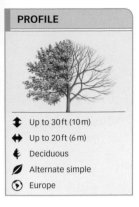

- Up to 30 ft (10 m)
- Up to 20 ft (6 m)
- Deciduous
- Alternate simple
- Europe

Succulent fruit
The fruit is an ovoid drupe with a thin skin, green at first turning red-purple. The succulent flesh inside ripens to a golden hue.

leaf margin bluntly serrated

leaf up to 3 in (7.5 cm) long

fruit up to 3 in (7.5 cm) long

fruit coated in a blue-white bloom

FEATURES

Bark Brown-gray, developing fissures with age

yellow stamens

Flowers Fragrant, borne in spring, before leaves appear

Prunus domestica (Rosaceae)

PLUM

Also known as garden plum, the original natural distribution of this small, broadly spreading tree is unknown. It may have developed as a hybrid between *Prunus spinosa* and *Prunus cerasifera*, both of which are native to the Caucasus region between Europe and Asia. Today, the plum is naturalized across much of Europe and is widely cultivated for its fruit throughout temperate regions of the world.

The tree has a broad and spreading crown. When young, its bark is smooth and has a slight sheen. The leaves are elliptic to obovate, up to 1 in (2.5 cm) wide, with a dull grass-green upper surface and slightly blue-green underside. Cherrylike white flowers, 1 in (2.5 cm) wide, are borne on bare branches. The sharp- to sweet-tasting, fleshy fruit grows up to 2 in (5 cm) wide. It contains a single flattened seed, known as a stone.

◌ **Sour cherry (*Prunus cerasus*)** Sour cherry has small, red, circular fruit, up to 1 in (2.5 cm) wide, which ripens to black. As indicated by its name, the fruit has a very sharp, acidic taste.

PROFILE

- Up to 22 ft (7 m)
- Up to 20 ft (6 m)
- Deciduous
- Alternate simple
- Japan

flower consists of five notched petals

Notched petals
White or pale pink flowers emerge from pink buds. The flowers are up to ¾ in (2 cm) wide and appear singly or in small clusters of two or three.

FEATURES

sharply incised leaf margin

leaf up to 3 in (7.5 cm) long, 2 in (5 cm) wide

Leaves Bronze-red in spring, fade to mid-green before turning bright red and gold

Prunus incisa (Rosaceae)

FUJI CHERRY

This hardy, slow-growing, deciduous Japanese cherry was first cultivated in 1910 from wild trees growing near Mount Fuji on Honshu, the main island of Japan. When introduced into the western world shortly afterward, it became hugely popular, and was widely planted in gardens and parks. Many trees at that time were grafted onto faster-growing rootstock, which resulted in incompatible and oversized growth at the trunk's base.

Young trees have an upright habit with ascending branches, while older trees are more wide-spreading. The gray-brown, smooth bark cracks in maturity. The flowers open before the leaves in early spring and are borne toward the ends of side twigs, off the main branch. The purple-black, egg-shaped fruit is up to ⅜ in (8 mm) long.

♧ Japanese apricot
This tree (*Prunus mume*) is much smaller than the Fuji cherry and its deeper pink flowers appear earlier in the year.

oval to
lanceolate leaf

flower up to
1 ½ in (4 cm) wide

Double-petaled flowers
The flowers are white, or white
flushed pink, double-petaled,
and without fragrance. They are
borne in short-stalked clusters
of two to five in spring.

PROFILE

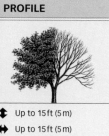

⬍ Up to 15 ft (5 m)

⬌ Up to 15 ft (5 m)

🍂 Deciduous

🗡 Alternate simple

◉ China, Japan

Prunus serrulata (Rosaceae)

HILL CHERRY

A small, flat-topped tree with spreading branches,
the hill cherry is grown as an ornamental tree in botanic
gardens and arboreta. It is believed to be the first
cultivated Japanese cherry to be introduced into the West,
arriving in Britain via Canton, China, in 1822. The species
is known to have been cultivated in Japan for at least
1,000 years; however, some botanists believe it may not
be a true species, but arose as a variant from an
unknown species of Chinese cherry.

The bark is gray-brown, with a slight sheen and
horizontal banding. The leaves are oval to lanceolate,
up to 5 in (13 cm) long and 2 in (5 cm) wide, with a long
tapered point and a finely serrated margin. Their upper
surface is bright green, while the underside is blue-
green. The fruit (seldom produced) is an inedible,
small, circular, red cherry.

☖ **Great white cherry (*Prunus* 'Taihaku')** This beautiful
Japanese cherry of garden origin is distinguished by its large,
dazzling white, single flowers, up to 3 in (8 cm) wide.

Spikelike racemes
Creamy white flowers are carried at the end of shoots in spreading to drooping, spikelike racemes in late spring or early summer.

smooth, glossy, dark green leaf upper surface

leaf up to 5 in (12 cm) long, 2 in (5 cm) wide

PROFILE

- Up to 80 ft (25 m)
- Up to 30 ft (10 m)
- Deciduous
- Alternate simple
- E. and C. North America

pale green leaf underside

flower up to ½ in (1 cm) wide

FEATURES

black ripe cherry

Fruit Round, edible cherry; red ripening to black

Prunus serotina (Rosaceae)

RUM CHERRY

Also known as black cherry, this medium-sized, broadly columnar tree is native to eastern and central North America, from Nova Scotia, Canada, to Mexico and Guatemala. It is called rum cherry because its fruit were once used by early North American settlers to flavor rum. Commonly found in woodland, it does not thrive as a single specimen in exposed regions.

The bark is dark gray and smooth, becoming horizontally fissured in maturity. It emits a distinctive, peppery odor when scratched. The leaves are elliptic to lanceolate, tapering to a point. They are finely toothed around the margin and have some hairs along the midrib. Rum cherry is a member of the bird cherry (p.233) group, and carries its flowers in racemes (pendulous clusters) up to 6 in (15 cm) long. Its fruit is a round cherry up to ½ in (1 cm) in diameter.

◊ **Kanzan flowering cherry (*Prunus serrulata* 'Kanzan')** This tree is distinguished from the rum cherry by its profusion of large pink flowers, smaller size, and spreading habit.

PROFILE

↕ Up to 30 ft (10 m)

↔ 30 ft (10 m)

🌿 Deciduous

🍃 Alternate simple

◐ China

flower borne on short stalk

pink flower has five petals

flower borne singly on shoot

Peach blossom
The flowers of the peach tree are pale to deep pink or red, and are produced before the leaves appear in early spring.

FEATURES

deep green upper surface of leaf

Leaves Shape varies from lanceolate to oblong-elliptic

Fruit Edible, rounded drupe, containing stone-like seed

Prunus persica (Rosaceae)

PEACH

This small, spreading tree has been cultivated for its fruit since time immemorial. The species name *persica* derives from an early European belief that this tree came from Persia (Iran). However, botanists now believe that the peach is probably native to China, and was introduced into Persia and much of the Mediterranean region, along the Silk Road, long before Christianity.

The bark is dark gray, smooth at first, becoming cracked into irregular plates in maturity. The leaves are up to 6 in (16 cm) long and 1¼ in (3 cm) wide. They have a slight sheen on the upper surface, with a pale green, matte underside. The flowers are produced either singly or, occasionally, in pairs. The edible, orange-yellow fruit grows up to 3 in (7 cm) wide, and has a velvety soft skin. It encloses yellow or whitish flesh, which becomes sweet and juicy when ripe. The flesh surrounds a large, deeply pitted seed, or stone. The stone has a pale brown woody husk, is oval in shape, and is up to 1 in (2.5 cm) long.

PROFILE

↕ 30–70 ft (10–20 m)

↔ Up to 50 ft (15 m)

🌿 Deciduous

🍂 Alternate simple

⊙ Japan

leaf sharply toothed at margin

Flowering cherry
Delicate pink flowers emerge in mid-spring against a backdrop of bronze-tinted foliage.

bronze-red young leaf

rose-pink flower, 1 1/4–1 1/2 in (3–4 cm) wide

flowers have notched petals

FEATURES

Bark Chestnut-brown with slight sheen; regularly marked with fawn-colored, linear lenticels ("breathing" pores)

long, slender leaf tip

Leaves Dramatic shades of orange in early to mid-fall

Prunus sargentii (Rosaceae)

SARGENT'S CHERRY

Named after Charles Sargent, first director of the Arnold Arboretum in Jamaica Plain, Massachusetts, this tree was introduced into Europe in 1890 by British plant collector Ernest H. Wilson. It is one of the larger Japanese flowering cherries, forming an open-rounded outline when mature.

At the base of the trunk on older trees, surface roots splay outward, sometimes for several yards, before disappearing underground. The shoots and buds are dark red-brown. Elliptic to obovate leaves end in a long, slender tip. They are bronze-colored when unfurling, fading to yellow and then dark green. Flowers are single, rose-pink, and borne profusely in clusters of four to five pairs in mid-spring, among emerging foliage.

◊ **Mount Fuji cherry**
This tree (*Prunus* 'Shirotae') has a short-trunked, wide-spreading habit and snow-white flowers.

PROFILE

- Up to 40 ft (12 m)
- 20 ft (6 m)
- Deciduous
- Alternate simple
- Europe

Hawthorn flowers
Strongly scented, creamy white flowers are produced in dense clusters on shoots around mid-spring.

dark green upper surface of leaf

pink anther

leaf cut deep, until near midrib

FEATURES

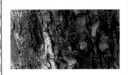

Bark Dull brown, smooth; becomes fissured with shallow, vertical cracks in maturity

fruit cluster on stalk

Fruit Berrylike, initially green, ripening to red

Crataegus monogyna (Rosaceae)

COMMON HAWTHORN

This slow-growing tree is often found growing in hedgerows and on open land, although it will grow on the edge of woodland too. Extremely hardy, it can withstand exposure, strong winds, and cold temperatures. It has been used in farming for centuries to shelter and enclose animal stock, particularly in upland areas. A regularly clipped hawthorn hedge makes an effective windbreak, with its strong mesh of small, thorny branches.

The bark is smooth at first, becoming fissured, with shallow, vertical, orange-colored cracks in maturity. Variously shaped leaves have paler undersides with some hairs in the vein axils. Creamy white flowers are followed by small berrylike fruit, known as haws. Initially green, haws ripen to red in fall and remain on the tree well into winter, when they become a valuable food source for migrating birds.

♧ **Oriental thorn (*Crataegus laciniata*)** This small ornamental tree has deeply cut, hairy leaves, which are darker green on the upper surface, and its fruit are more yellowish red.

Glossy red fruit
This tree bears oval, red fruit, or haws, which remain on the tree till late fall, sometimes winter, before being eaten by birds.

shallow-lobed leaf with toothed margin

leaf up to 2 in (5 cm) long and wide

fruit up to 1 in (2.5 cm) long

PROFILE

↕ Up to 30 ft (10 m)

↔ Up to 25 ft (8 m)

🍂 Deciduous

🌿 Alternate simple

◉ N.W. and C. Europe

FEATURES

Flowers Creamy white, pungently scented, with five petals and pink-orange anthers

mottled leaf

Variant Young leaves of *Crataegus laevigata* 'Gireoudii' are mottled cream and pink

Crataegus laevigata (Rosaceae)

MIDLAND HAWTHORN

This hardy, broadly spreading tree takes its name from an area in central England, where it grows in profusion, and is an accepted indicator species for evidence of ancient woodland. It is native to northwest and central Europe, where it is commonly found in hedgerows and on woodland edges. It differs from common hawthorn (p.243) in its leaves, which are less deeply lobed, and its flowers, which have two styles (female part of flower) as opposed to one in the common hawthorn.

The bark is gray-brown, smooth at first, cracking into small plates in maturity. Leaves are ovate to obovate, glossy dark green on the upper surface, paler on the underside. Thorns, up to ¾ in (2 cm) long, are sporadically borne on tips of twigs. Flowers appear in small but dense clusters in late spring. These are followed by small, oval fruit, each containing two hard seeds.

♧ ***Crataegus dsungarica*** This small hawthorn is distinguished from the midland hawthorn by its profusion of thorns, larger flowers, and fruit, which turn purple-black rather than red.

Fruit clusters
Fleshy and rounded, edible, bright red fruit hang in clusters from thin stalks.

obovate leaf

sharp spine up to 3in (8cm) long

serrated leaf margin

fruit up to ½in (1cm) wide

PROFILE

- Up to 25ft (8m)
- Up to 25ft (8m)
- Deciduous
- Opposite simple
- E. and S. North America

FEATURES

Bark Dark brown and scaly; peels to reveal fresh orange, new bark beneath

Flowers Creamy white, borne in dense, rounded heads in late spring

Crataegus crus-galli (Rosaceae)

COCKSPUR HAWTHORN

A small, hardy, broadly spreading tree, the cockspur hawthorn owes its name to its ferociously sharp spines. This tree is native to eastern and southern North America, where it is found growing wild on the edge of woodland from Quebec, Canada, south to Florida, and west to Texas. It was introduced into the Old World in 1691 and is now commonly planted in towns and cities, alongside streets and in parking lots.

The leaves are glossy dark green on the upper surface; pale green on the underside. Up to 4in (10cm) long and 2in (5cm) wide, they are rounded at the tip and tapered at the base. In fall, they turn a bright marmalade-orange color before being shed. The shoots bear sharp spines up to 4in (10cm) long. The white flowers have yellow-pink anthers up to ½in (1.5cm) long. Fruit persist on the tree long after leaf fall.

◊ **Azarole (*Crataegus azarolus*)** This tree has rhomboid leaves, which are wedge-shaped at the base. It has large white flowers with purple anthers, and pale yellow to orange fruit.

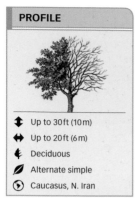
leaf blue-green on underside

leaf deep green on upper surface

fruit up to 1½in (4cm) wide

Fruit and leaves
The fruit are small, rounded apples, green becoming yellow, sometimes flushed with red. Elliptic to ovate leaves are 1½in (4cm) long.

FEATURES

Bark Brown, smooth; becomes fissured when relatively young

Flowers Slightly fragrant, up to 1in (2.5cm) wide

Malus sylvestris (Rosaceae)

COMMON CRAB APPLE

This small, broadly spreading tree is probably best known as one of the parents of the cultivated apple (p.249). It is also known as the sour apple because of the acidic taste of its small fruit, which are inedible when tasted raw. These fruit can, however, be made into a crab apple jelly.

This tree produces an uneven, low-domed crown of dense, twisting branches, normally weighted to one side. Its elliptic to ovate leaves are slightly rounded at the base, with some tufts of hair along the leaf veins. The leaf margin is finely, but bluntly, serrated. The flowers of the common crab apple emerge white, or white flushed with pink, in spring. They are carried in loose clusters on short spurs. The flowers develop into a fruit with creamy white, juicy flesh inside, which surrounds a central core containing dark brown, almost black, oval seeds.

◊ **Snowy mespil (*Amelanchier lamarckii*)** This tree has oval to oblong, coppery pink leaves in winter, and lax racemes of starlike white flowers in spring. It bears small, round, black-purple berries.

Tapering leaves
The leaves are elliptic to ovate, and taper to a fine point with some hairs along the midrib and main veins.

bright grass-green leaf upper surface

finely toothed leaf margin

Deciduous

Alternate simple

C. China

pale green leaf underside

small, rounded fruit, up to ½in (1cm) wide

fruit borne on long, slender stalk

FEATURES

flower 2in (5cm) wide

Flowers Fragrant, white with pink flush; with five broad, overlapping petals

Malus hupehensis (Rosaceae)

HUPEH CRAB

This broadly spreading tree is one of the most beautiful of all small deciduous trees. It reaches up to 40ft (12m) in its native land among the mountains of Hupeh Province in central China. In cultivation, however, it is normally smaller with stiff, ascending branches. In China, its leaves are used to make a drink called red tea. It was introduced into western cultivation in 1900 by the English plant collector Ernest H. Wilson.

Its bark is purple-brown with irregular plates that flake to reveal fresh, orange-brown bark beneath. Its flowers are pink as buds, but open white with a rose flush in spring. In a good year, they are borne in such profusion that the whole tree canopy seems to be covered in blossoms. Its fruit is greenish yellow, ripening to a small, slightly flattened, red "apple," pendulously borne in clusters of three or four.

♤ **Allegheny serviceberry (*Amelanchier laevis*)** This tree has soft, lanceolate leaves, which emerge a bronze-pink color amidst a profusion of starlike flowers in early spring.

long petiole
(leaf stalk)

fruit up to
¾ in (2 cm) wide

sharply toothed
leaf margin

leaf up to 3 in (8 cm) long,
1 ½ in (3.5 cm) wide

PROFILE

↕ Up to 50 ft (15 m)

↔ Up to 40 ft (12 m)

🌿 Deciduous

🍃 Alternate simple

📍 Siberia, N.E. Asia,
N. China

stiff, red
pedicel

Fruit and leaves
The fruit is yellow-green at first,
ripening to red. The lanceolate
leaves are rich green on the upper
surface with a paler underside.

FEATURES

elliptic-shaped
petal

Flowers White; up to 1 ½ in
(3.5 cm) wide

Malus baccata (Rosaceae)

SIBERIAN CRAB

Native to Siberia, as well as other parts of northeast
Asia and northern China, this small to medium-sized tree
is hardy, round-crowned, and slightly pendulous. It was
first cultivated as an ornamental specimen outside this
region in 1784. Today it is found in parks, gardens, and
arboreta throughout western Europe and North America.

The bark is smooth at first, flaking into irregular
patches with age to reveal a lighter, gray-brown bark
beneath. The leaves are somewhat similar to those
of the common pear (p.250). They are attached to
the shoots by a 2 in- (5 cm-) long, slender petiole (leaf
stalk). Fragrant flowers have widely parted petals
surrounding a central cluster of yellow anthers. They
are borne in mid-spring in groups of five, each on a
stiff pedicel up to 2 ½ in (6 cm) long. The globular fruit
remains on the tree well into winter.

🌼 **Pillar crab (*Malus tschonoskii*)** Unlike the Siberian crab, this
tree is erect with long, upright branches. Additionally, the leaves are
coated with gray-white hairs on the underside.

PROFILE

↕ Up to 30 ft (10 m)

↔ Up to 20 ft (6 m)

🌿 Deciduous

🌿 Alternate simple

📍 Originated in W. Asia; exists today in cultivation worldwide

Variable fruit
The fruit is variable in color and taste, with creamy white flesh surrounding five chambers (carpels), each bearing one to three seeds.

leaf up to 5 in (12 cm) long, 2½ in (6 cm) wide

fruit up to 3½ in (9 cm) wide

FEATURES

Bark Gray-brown, smooth; fissures into irregular-shaped plates with age

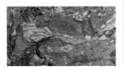

Flowers Five-petaled, fragrant; white with pink tinge that fades over time

Malus domestica (Rosaceae)

CULTIVATED APPLE

Also known as orchard apple, this is a small, broadly spreading tree grown throughout the world for its edible fruit. It was originally cultivated in western Asia and probably arose as a hybrid of several *Malus* species including *M. sylvestris* and *M. sieversii*. Today there are more than 7,500 known cultivars, each one producing a slightly different fruit, flower, form, or size.

Its bark peels in thin flakes to lift and reveal lighter brown, new bark beneath. The leaves are deep green on the upper surface and paler on the underside. They have a pointed tip, serrated margin, and a slightly downy underside. They are borne on a 2 in- (5 cm-) long petiole. The fragrant flowers are up to 1½ in (3.5 cm) wide. The juicy, edible fruit matures in late summer.

Spreading tree
The cultivated apple tree is broadly spreading, with a rounded head and hairy, young shoots.

PROFILE

- ⬍ Up to 50 ft (15 m)
- ⬌ Up to 40 ft (12 m)
- Deciduous
- Alternate simple
- Originated in W. Asia; exists today in cultivation worldwide

ovate to elliptic leaf, rounded at base

leaf up to 4 in (10 cm) long, 2 in (5 cm) wide

fruit typically 4 in (10 cm) long

Variable fruit
Pears vary in shape and color depending on the cultivar: from rounded to pear-shaped; green, speckled brown, russet, or red.

FEATURES

Flowers White, five-petaled, with purple-red anthers; borne in clusters in spring

Pyrus communis (Rosaceae)

COMMON PEAR

This tree is believed to be a hybrid that originated in western Asia over 2,000 years ago, like the orchard apple (p.249). It has been cultivated throughout the western world for centuries, with several popular cultivars having been raised at Versailles, France, in the 17th century. Today, there are over 1,000 different cultivars available commercially, each one with slightly different form, size, fruit, or flowers.

The common pear is a broadly columnar, upright, small to medium-sized tree. It has distinctive spur shoots, which occasionally develop into spines and give the crown a slightly angular appearance. The bark is dark gray, fissuring into small irregular plates in maturity. Its leaves are glossy dark green on the upper surface, paler on the underside. The edible fruit is sweet and juicy, with creamy white flesh, which surrounds small, hard, brown seeds.

♤ **Wild Himalayan pear (*Pyrus pashia*)** This can be distinguished from the common pear by its tangled mass of spiny branches, and its smaller brown fruit, which are speckled with cream dots.

leaf twists along its length

purple anther

leaf up to 4 in (10 cm) long, less than 1½ in (4 cm) wide

flower up to 1 in (2.5 cm) wide

PROFILE

⬍ Up to 30 ft (10 m)

⬌ Up to 22 ft (7 m)

🍂 Deciduous

🌿 Alternate simple

◉ Russia, Caucasus region, W. Asia

Flowers in clusters
Creamy white flowers with purple anthers are borne in widely spaced clusters along the branches in spring.

FEATURES

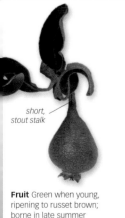

short, stout stalk

Fruit Green when young, ripening to russet brown; borne in late summer

Pyrus salicifolia (Rosaceae)

WILLOW-LEAVED PEAR

This small, broadly weeping tree is also known as the weeping silver-leaved pear. The most ornamental of pears, it is native to Russia, Turkey, and northern Iraq. It was introduced in the West by the German botanist and explorer P. S. Pallas, who discovered it in 1780. Today it is commonly planted in gardens and parks. It is not a long-lived tree.

Its bark is pale gray, becoming vertically fissured in maturity. The leaves are narrowly lanceolate, usually untoothed, and tapering at both ends. They also display a twist along their length. Young leaves are covered with a silvery white down, which extends to the pendulous shoots. Mature leaves are sage green and smooth. The five-petaled flowers appear on short shoots in spring. The fruit is a small, hard pear, up to 1¼ in (3 cm) long. Hairy when young, it turns smooth with age.

◊ **Sand pear (*Pyrus pyrifolia*)** Also called the Chinese pear or Asian pear, this tree has ovate, toothed, glossy green leaves, up to 5 in (12 cm) long, which turn deep red in fall.

Sharply toothed leaflets
The pinnate leaves, up to 8 in (20 cm) long, carry up to 15 leaflets that are sharply toothed along the margin and taper to a point.

leaflet up to 2 in (5 cm) long

fruit borne in pendulous clusters

round, red berry up to ½ in (1 cm) wide

PROFILE

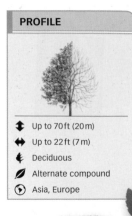

↕ Up to 70 ft (20 m)

↔ Up to 22 ft (7 m)

🌢 Deciduous

🗲 Alternate compound

◉ Asia, Europe

FEATURES

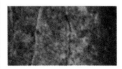

Bark Silver-gray, smooth, shiny even in maturity

creamy white flower, up to ½ in (1 cm) wide

Flowers Fragrant, borne in dense clusters, up to 6 in (15 cm) wide, in late spring

Sorbus aucuparia (Rosaceae)

ROWAN

This broadly conical, deciduous tree is native to the whole of Asia and Europe. In mountainous regions of northern Europe, it can grow up to 3,280 ft (1,000 m) above sea level—hence its other name mountain ash. It is a particularly hardy tree and can withstand exposure better than most members of the *Sorbus* genus. Its berries are an important food source for birds. In some regions of Europe, a rowan tree planted outside the house is considered protection against evil spirits.

The leaves have a fresh green upper surface and a blue-green underside. In fall, they turn red or yellow before being shed. The fruit is borne toward the end of summer on stalks up to ¾ in (2 cm) long and remains on the tree until eaten by birds.

◌ **Japanese rowan**
This tree (*Sorbus commixta*) has larger leaflets than the mountain ash, and red, long-pointed winter buds.

- 50 ft (15 m)
- Up to 70 ft (20 m)
- Deciduous
- Alternate simple
- N., W., and C. Europe

eaves and berries
he leaves are ovate
o elliptic with irregular,
hallow serrations around
he margins. Clusters of red
erries are borne in summer.

red, oval to rounded berry

FEATURES

Flowers Creamy white, fragrant; up to ½ in (1 cm) wide

Sorbus aria (Rosaceae)

WHITEBEAM

Native to north, west, and central Europe, the whitebeam is a small to medium-sized tree of calcareous uplands. It thrives on thin chalk and limestone soils. It is sometimes referred to as the weather tree, due to the fact that in windy conditions—often the precursor to low pressure—the wind flips the leaves over, revealing their white underside. This is taken by some as a sign that rains are imminent.

The leaves are up to 3½ in (9 cm) long and 2 in (5 cm) wide. Initially pale green, they turn matte green on the upper surface and white on the underside. A dense covering of silver-white hairs extends down the leaf stalks. The flowers are clustered in flattened heads, each up to 3 in (8 cm) wide. The fruit is a rough, red berry with fawn speckles.

Rounded habit
Conical when young, the whitebeam matures into a broadly columnar tree with a rounded habit.

PROFILE

Up to 50 ft (15 m)

Up to 40 ft (12 m)

Deciduous

Alternate simple

N.W. Europe

Swedish whitebeam flowers
The dull white flowers of this tree are individually small, but they are borne in large, flattened clusters up to 5 in (12.5 cm) wide.

yellowish white stamens

flower ¾ in (2 cm) wide, with five petals

FEATURES

ovate leaf, with toothed margin

Fruit Deep red, oval berry, up to ½ in (1 cm) long

Sorbus intermedia (Rosaceae)

SWEDISH WHITEBEAM

A medium-sized tree, the Swedish whitebeam is native to northwest Europe including much of southern Scandinavia. It is an extremely hardy tree that thrives in exposed conditions, and can tolerate localized pollution—hence its popularity as an urban street tree. It is also used to restore land previously affected by industrial pollution as it can grow on impoverished soils.

The bark is gray and shiny at first, cracking into uneven flakes on maturity. The irregularly lobed leaves are dark green with a slight sheen on the upper surface and pale green on the underside, with a dense covering of short, white hairs. The dull white flowers are followed by fruit at the end of summer.

Arching branches
The Swedish whitebeam has a broad crown. Its arching branches become pendulous toward the tips.

deeply cut lobed leaf,
up to 4 in (10 cm)
long and wide

creamy
white flower

↕ Up to 70 ft (20 m)

↔ Up to 50 ft (15 m)

🌿 Deciduous

🍃 Alternate simple

◉ N. Europe, N. Africa

flowers borne in
clusters, up to 4 in
(10 cm) wide

Flowers and leaves
Small flowers appear in late spring
to early summer. Leaves are dark
green on the upper surface, and
pale green on the underside.

FEATURES

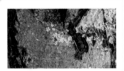

Bark Gray-brown; develops
irregular plates with age

Fruit Oval-shaped, warty berry,
up to ½ in (1.5 cm) long

Sorbus torminalis (Rosaceae)

CHECKER TREE

Also known as the wild service tree, the checker
tree is easily distinguished from the members of the
genus *Sorbus* by its stiff, maple- or plane-shaped leaves.
Before the introduction of hops, its fruit were used to
flavor beer—hence the frequently found English pub
name: "The Chequers."

Smooth at first, the gray-brown bark cracks into
irregular plates, flaking off to reveal fresh, rust-brown
bark beneath. Winter leaf buds
are distinctly olive-green. The
leaves are deeply cut into
sharply toothed lobes, and fixed
alternately on olive-brown
shoots by yellow-green stalks
up to 2 in (5 cm) long. In fall,
the leaves turn yellow-brown
before being shed. Erect
clusters of russet brown fruit
follow in early fall.

Broadly columnar tree
The medium-sized
checker tree has a broadly
columnar habit and
distinctive leaves.

PROFILE

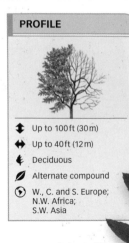

⭥ Up to 100 ft (30 m)

⭤ Up to 40 ft (12 m)

❦ Deciduous

🌿 Alternate compound

⊕ W., C. and S. Europe;
N.W. Africa;
S.W. Asia

*fruit up to 1½ in
(4 cm) in diameter*

*yellow-green leaflet,
up to 2½ in (6 cm) long,
½ in (1 cm) wide*

Leaves and fruit
The pinnate leaves are up to 10 in
(25 cm) long and consist of 13 to 21
oblong leaflets. The greenish-brown
fruit is sometimes flushed red.

*serrated
margin on top
half of leaflet*

FEATURES

*flower up
to ¾ in
(1.8 cm) wide*

Flowers Creamy white, with
five petals

Sorbus domestica (Rosaceae)

SERVICE TREE

Although it is larger in size, at first glance the service
tree resembles the rowan (p.252). This broadly columnar,
medium- to large-sized tree may occasionally be up to
100 ft (30 m) tall. It is native to western, central, and
southern Europe; northwest Africa; and southwest Asia.
However, it is rare in some of these regions; for example,
it is found growing wild in only a few locations in the
British Isles. It is sometimes referred to as the true
service tree to distinguish it from the wild service tree
(p.255), another name for the checker tree.

The brown bark is smooth on young trees, but
becomes fissured and flaky in maturity. Winter buds
are green and sticky. In late spring, flowers are densely
borne in flattened clusters known as corymbs, which are
up to 5½ in (14 cm) wide. The fruit is either apple-shaped
(var. *pomifera*) or pear-shaped (var. *pyrifera*).

◊ **Apple-fruited service tree (*Sorbus domestica* var.
pomifera)** This tree can be distinguished from the service tree
by its redder, more rounded, apple-shaped fruit.

leaflet with
pointed tip

sharply toothed margin
on top half of leaflet

flower up to ¼ in
(7 mm) wide

Spring blooms
The white, five-petaled flowers
of the Hupeh rowan bloom in
late spring, in dense clusters up
to 4 in (10 cm) wide.

Fruit Rounded, white to
pink in color, up to ½ in
(1 cm) wide

Sorbus hupehensis (Rosaceae)

HUPEH ROWAN

Also known as the Chinese mountain ash, this small,
broadly columnar, deciduous tree is native to central
and western China. Introduced into western cultivation
in 1910, it is now widely grown in parks, gardens, and
arboreta throughout the temperate world. It is a popular
ornamental tree in Europe and North America, where it
produces a compact crown of slender, ascending,
purple-brown branches.

The bark is gray-brown to purplish brown. It is
smooth when young, with a slight sheen, becoming
flaky in maturity. The leaves are ash-like, pinnate, and,
unlike most other rowans, blue-green in color. Each
pinnate leaf is made up of 11 to 17 oval-shaped leaflets
that turn brilliant red in fall. The flowers are slightly
fragrant. The fruit is borne in loose, drooping bunches
in fall, and persists into winter.

♀ **Vilmorin's rowan (*Sorbus vilmorinii*)** This tree has smaller
pinnate leaves, each with 19 to 25 oblong to elliptic, round-tipped
leaflets. The fruit is smaller and deep pink.

narrowly elliptic leaflet,
up to 2 in (5 cm) long

leaflet tapers
to a point

rounded fruit,
up to ½ in
(1 cm) wide

fruit borne
on pendulous,
red-stalked panicle

Foliage and fruit
The leaves are pinnate and made
up of 15 to 19 leaflets. Fruit ripen
from pink to white, and look like
waxy marbles.

FEATURES

Flowers Five-petaled, soft
pink, carried in dense clusters
in late spring

Sorbus cashmiriana (Rosaceae)

KASHMIR ROWAN

This small, broadly conical tree was first described
from specimens collected in Kashmir, northern India,
in 1901. However, it was not cultivated outside its
natural range of the western Himalayas until the
1930s. Since then it has been widely planted
throughout the temperate world and is a prized
addition to gardens and arboreta.

It has gray, sometimes red-gray, bark that remains
relatively smooth even in maturity. The winter leaf buds
are conical and distinctly pink-red. Leaves are pinnate,
up to 6 in (15 cm) long. Each leaflet is deep green on the
upper surface and pale green on the underside, with a
fine covering of silver-gray hairs. Leaflets are sharply
serrated along the margins. Slightly fragrant, light pink
flowers up to ½ in (1.5 cm) wide are carried in lax
corymbs up to 4 in (10 cm) in width in late spring. These
are followed by berrylike, pink-colored fruit, which turn
to white when ripe. They persist on the tree long after
the leaves have fallen in early winter.

slender-pointed
leaflet

scarlet fruit,
each up to ½ in
(1 cm) wide

leaflet up to 4 in (10 cm)
long, 2 in (5 cm) wide

Fruit and leaves
Large, lax, rounded heads of
fruit up to 6 in (15 cm) wide grow
alongside grass-green leaves
with gray-green undersides.

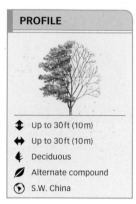

PROFILE

↕ Up to 30 ft (10 m)

↔ Up to 30 ft (10 m)

↯ Deciduous

⬦ Alternate compound

☉ S.W. China

FEATURES

Leaves Rich red in fall before
being shed

red, conical
bud

Buds Sticky;
appear in
winter

Sorbus sargentiana (Rosaceae)

SARGENT'S ROWAN

This medium-sized rowan is native to the mountains
of Sichuan in southwest China. It was collected by British
plant collector Ernest H. Wilson in 1908 for the Arnold
Arboretum in Jamaica Plain, Massachusetts, and was
named after the arboretum's director, Charles Sprague
Sargent. Known for its attractive fruit and fall color, this
tree has rigid, widely spreading branches.

The bark is plum-brown and smooth, even in maturity.
The leaves are pinnate and up to 14 in (35 cm) long. They
consist of 7 to 11 leaflets that have fine hairs and
serrations around the upper half of the margin. The leaf
stalks are bright crimson-red. The large winter buds are
similar to those of the horse chestnut (p.281), but more
conical and red. In early summer, small, creamy white
flowers are borne in large, dense corymbs (flat-topped
clusters of flowers) up to 8 in (20 cm) wide.

♤ **Chinese mountain ash (*Sorbus scalaris*)** Unlike Sargent's
rowan, this tree has leaves with more than 30 small leaflets. They
have a glossy green upper surface and gray hairs on the underside.

PROFILE

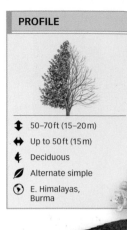

↕ 50–70 ft (15–20 m)

↔ Up to 50 ft (15 m)

🌢 Deciduous

⬙ Alternate simple

◉ E. Himalayas, Burma

creamy white flower, up to ½ in (1 cm) wide

serrated margin around top half of leaf

deep green leaf, 6 in (15 cm) long

Corymbs and leaves
The flowers are borne in dense clusters (corymbs) on thick, woolly stalks. Leaves are large and oval to almost round in shape.

FEATURES

small white hairs on underside

Leaves Green-white on underside; up to 6 in (15 cm) long and wide

Sorbus thibetica (Rosaceae)

TIBETAN WHITEBEAM

This silvery, white-leaved whitebeam was originally collected from the Himalayas in the 1920s. It is rarely seen in cultivation in its true form, and is mainly represented by the cultivar 'John Mitchell.' This cultivar is named after the one-time curator of Westonbirt Arboretum in Gloucestershire, England, who selected this form and planted it at Westonbirt in 1938.

A large-sized rowan, it has wide-spreading branches and a broad, rounded head of dense foliage. Its silver-gray bark is smooth and shiny at first but becomes darker and lightly fissured in maturity. The pungently scented flowers have purple-orange anthers and are similar to those of the common hawthorn (p.243). They are carried densely on corymbs (flat-topped flower clusters) up to 4 in (10 cm) wide. The fruit is apple-shaped, up to ½ in (1 cm) wide, green ripening to russet-brown.

♧ **Sorbus megalocarpa** This tree is distinguished by its broadly obovate, coarsely toothed leaves with little hairs on the underside, and its large, marble-sized, brown, speckled fruit.

PROFILE

↕ Up to 20 ft (6 m)

↔ Up to 20 ft (6 m)

🌢 Deciduous

🍃 Alternate simple

🧭 S.E. Europe,
S.W. Asia

Roselike flowers
The white flowers have five
well-spaced petals around
a central cluster of purple
anthers and golden stamens.

*bright green leaf, up
to 6 in (15 cm) long*

*flower up to
2 in (5 cm)
wide*

*minutely toothed
leaf margin*

FEATURES

Fruit Russet brown; shaped
like a flattened pear

Mespilus germanica (Rosaceae)

MEDLAR

This small, broadly spreading, deciduous tree
is native to southwest Asia and southeast Europe.
It has long been cultivated for its fruit, which is
edible and tasty, though considered an acquired
taste by some. This is probably due to the need
to "blet" the fruit before eating. This means
allowing the fruit to overripen until it becomes
like a soft pear. Once bletted, the fruit tastes like
cinnamon-flavored stewed apple.

The bark is smooth and dull brown, becoming
fissured and flaking with age. The elliptic to lanceolate,
stalkless leaves are faintly hairy on both sides. The
flowers are borne singly at the ends of shoots in early
summer. The fruit is up to 2 in (5 cm) long and 1¼ in
(3 cm) wide and has an open top surrounded by a
calyx, similar to a rose hip—the fruit of the rose.

🗘 **Quince *(Cydonia oblonga)*** This tree is distinguished by its
broadly elliptic to rounded leaves, green shoots, pale pink flowers
and large, green-yellow, pear-shaped fruit.

PROFILE

- Up to 80 ft (25 m)
- 50 ft (15 m)
- Deciduous
- Alternate compound
- United States

flowers in pendulous raceme, up to 8 in (20 cm) long

reddish brown calyx

smooth leaflet margin

leaflet up to 2 in (5 cm) long

ovate to elliptic leaflet, grass green on upper surface

Leaves and flowers
Pinnate leaves, up to 12 in (30 cm) long, have 11 to 21 lax leaflets. Pealike, fragrant white flowers are borne in early summer.

FEATURES

Fruit Reddish to dark brown, hanging pod bears black seeds

Robinia pseudoacacia (Fabaceae)

BLACK LOCUST

Native to the Appalachian Mountains, from Pennsylvania to Georgia, this large, broadly columnar tree is now naturalized throughout the United States. The black locust was one of the first American trees to be introduced into Europe, in 1601.

The smooth, gray-brown bark becomes fissured and vertically ridged with age. The branch wood often breaks to form scars on old trees. On mature trees, suckers from the roots can rise up to 30 ft (10 m) away from the trunk. The sharply pointed leaves are blue-green on the underside, and may have two spines at the base. The fruit is an oblong pod up to 4 in (10 cm) long. It splits open in late summer to reveal black seeds.

♧ **Golden-leaved robinia**
Unlike the black locust, this cultivated variety (*Robinia pseudoacacia* 'Frisia') has bright golden yellow leaves and a smaller stature.

↕ Up to 100 ft (30 m)

↔ Up to 70 ft (20 m)

🌡 Deciduous

🖊 Alternate compound

◉ N. China

glossy green upper surface of leaflet

untoothed leaflet margin

leaflet up to 2 in (5 cm) long, tapers to narrow point

Ovate leaflets
The pinnate leaves are alternately arranged on blue-green shoots. Up to 10 in (25 cm) long, each leaf has 9 to 17 opposite leaflets.

FEATURES

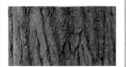

Bark Light gray, smooth; develops regular vertical fissures in maturity

Flowers White, pealike, borne on pendulous panicles

Sophora japonica (Fabaceae)

SCHOLAR TREE

Despite its alternative name, Japanese pagoda tree, this broadly columnar tree is native to northern China and not Japan. However, it has been widely cultivated in Japan for centuries, particularly in temple gardens in and around Kyoto. It was introduced into western cultivation in 1753. When grown from seed, it may take up to 30 years to produce flowers.

The bark is reminiscent of that of the common ash (p.300). The leaves are blue-green and slightly hairy on the underside, and emerge from buds only in early summer. Fragrant white flowers are carried on terminal, pendulous, widespread panicles, up to 10 in (25 cm) long, in late summer. The flowers are more abundant after hot summers. The fruit is a pod, green ripening to brown, up to 3 in (7.5 cm) long. It contains up to six seeds.

🌢 **Kowhai (*Sophora tetraptera*)** This small, evergreen tree has pinnate leaves, with 20 to 40 ovate to elliptic leaflets, each up to 1 ¼ in (3 cm) long, and yellow, tubular flowers.

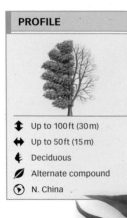

↕ Up to 100 ft (30 m)

↔ Up to 50 ft (15 m)

🍂 Deciduous

🌿 Alternate compound

◐ N. China

large leaflet, up to 6 in (15 cm) long

dark green upper surface of leaflet

Pinnate leaves
The pinnate leaves are up to 30 in (75 cm) long. They consist of 11 to 19 leaflets held on a bright red petiole (stalk), up to ½ in (1.5 cm) long.

slightly crinkled, oblong to lanceolate leaflet

FEATURES

Bark Slate-gray, relatively smooth; develops vertical, light patterning in maturity

Fruit Winged seed; yellow-brown, tinged with red when ripe

Ailanthus altissima (Simaroubaceae)

TREE OF HEAVEN

One of the fastest growing of all temperate Asian trees, the tree of heaven was introduced into western cultivation in 1751. This large, broadly columnar tree produces extensive suckers from its roots, and has been known to lift sidewalks and crack pavement up to 50 ft (15 m) from the trunk.

Leaves consist of 11 to 19 leaflets, which have some serration near the base. Borne in late spring, they are bright red at first, developing a deep green upper surface and a pale blue-green underside later. Yellow-green male and female flowers are borne in large panicles on separate trees in summer. Female trees bear winged seeds in ash-like, orange-red bunches, up to 12 in (30 cm) long.

Pollution-resistant tree
The tree of heaven is widely planted in towns and cities for its ability to withstand atmospheric pollution.

Bipinnate foliage
In summer, this tree bears large, bipinnate leaves, 3 ft (1 m) long, consisting of up to 14 leaflets.

leaflet up to 3 in (7.5 cm) long

ovate, untoothed leaflet

PROFILE

↕ Up to 70 ft (20 m)
↔ 25–30 ft (8–10 m)
⚘ Deciduous
⬤ Alternate compound
◉ C. and E. United States

FEATURES

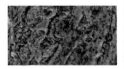

Bark Dark gray, red-tinged, with deep fissures

Fruit Seed capsule, glaucous-green when unripe

Gymnocladus dioicus (Fabaceae)

KENTUCKY COFFEE TREE

This large, broadly columnar, sometimes spreading tree is native to the central and eastern United States. Early European settlers would roast and use its seeds to make a substitute coffee-like drink—hence the reference to coffee in its name.

The Kentucky coffee tree is rather sparse looking, with branches that remain bare until toward their tips. The leaves are produced late in the season, and for a while in spring the tree shows no signs of life because the leaf buds are small and so easily missed. The leaflets are bronze when young, developing a dark green upper surface and a blue-green underside in summer. They turn golden yellow in fall. Greenish white, fragrant flowers, up to 1 in (2.5 cm) wide, are borne in conical panicles, up to 4 in (10 cm) long. The flowers are followed by broad-bean-like seed pods, up to 10 in (25 cm) long and 2 in (5 cm) wide. Glaucous-green at first, they ripen to red-brown. The seed pods contain up to nine seeds set within a sweet-tasting pulp.

PROFILE

↕ 70–100 ft (20–30 m)

↔ Up to 70 ft (20 m)

⚘ Deciduous

🌿 Alternate compound

◔ C. and E. US

leaflet tapers to a broad tip

Honey locust leaves
The leaves are pinnate, sometimes bipinnate (where the leaflets themselves are pinnate). Each leaf has up to 36 oblong-lanceolate leaflets.

glossy bright green upper surface of leaflet

leaflet up to 1½ in (4 cm) long

FEATURES

Flowers Small, yellow-green, up to ½ in (1 cm) wide; borne in racemes up to 3 in (7.5 cm) long in early to midsummer

Gleditsia triacanthos (Fabaceae)

HONEY LOCUST

This broadly spreading tree is best identified by its ferocious thorns, which cover its trunk and main branches. Its native habitat extends from New England to Texas, and it has been cultivated in Europe since its introduction in 1700. It is widely planted in city streets because of its tolerance to heat, dust, drought, and airborne pollutants. Unlike other members of the pea family, the honey locust does not produce the typical "wing and keel" pea flower.

The bark is gray-brown, smooth when young, developing clusters of sharp spines, up to 3 in (7.5 cm) long, in maturity. Leaves are frond-like in appearance and lack a terminal leaflet. The leaflets have slight serrations along the margins. The shoots are green and may carry smaller spines next to each leaf bud. The seed-bearing fruit is a large, twisted, pendulous pod up to 18 in (45 cm) long. Green at first, it ripens to olive-brown.

Yellow-wood leaves
The pinnate leaves are 8–12 in (20–30 cm) long, with 5 to 11 broadly ovate to oval leaflets.

leaflet up to 5 in (13 cm) long, 3 in (7 cm) wide

dark green upper surface of leaflet

PROFILE

⬍ Up to 50 ft (15 m)

⬌ Up to 70 ft (20 m)

🌢 Deciduous

⬧ Alternate compound

◔ S.E. United States

FEATURES

Bark Smooth gray to light brown in color

untoothed leaflet margin

Leaves Bright clear yellow in fall

Cladrastis kentukea (Fabaceae)

YELLOW-WOOD

Also known as virgilia, this broadly columnar tree was introduced into Europe in 1812. It is a small to medium-sized tree with a broad and rounded crown. It takes its name from its yellow heartwood, which is used to make furniture, gun stocks, and decorative items.

The shoots are green and downy when young, turning yellow-green, and then red-brown, with age. The leaflets are dark green on the upper surface, while the underside is pale green and covered with fine hairs. White, mildly fragrant flowers are produced in early summer. They are borne in wisteria-like, pendulous racemes. These are followed by fruit, which are smooth pods, green ripening to brown. They contain two to six dark brown seeds.

Low-branched tree
Yellow-wood normally has low branching and a wide crown of upright, spreading branches.

PROFILE

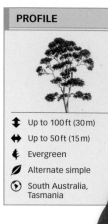

↕ Up to 100 ft (30 m)

↔ Up to 50 ft (15 m)

✦ Evergreen

∅ Alternate simple

⊙ South Australia, Tasmania

oval, woody seed capsule, up to ¼ in (5 mm) long

lanceolate, blue-green mature leaf

mature leaf, 5 in (12 cm) long, 1½ in (4 cm) wide

White flowers
In summer, flowers with numerous, creamy yellow stamens are borne in clusters of three in the leaf axils.

FEATURES

Bark Mottled gray-green and copper; flakes in long strips to reveal cream surface below

juvenile blue-green leaf

Leaves
Glaucous silver-blue, stalkless; carried in opposite pairs when young

Eucalyptus gunnii (Myrtaceae)

CIDER GUM

This large, broadly columnar, fast-growing evergreen is native to Tasmania, where it grows in moist mountain forests up to 4,265 ft (1,300 m) above sea level. One of the hardiest of all *Eucalyptus* species, it is widely planted in temperate regions as an ornamental tree and for timber. The attractive juvenile foliage is highly valued by flower arrangers and florists. In order to maintain this foliage, trees are regularly coppiced (trimmed down to stumps), to stimulate the growth of new shoots.

The juvenile leaves are rounded and up to 1½ in (4 cm) long. They are virtually stalkless, and are carried in opposite pairs on pink, bloomy shoots. Mature leaves are alternate and hang down from limp branchlets. The flowers are slightly fragrant, and after pollination, develop into dry, gray seed capsules that are open at one end.

◌ **Urn gum (*Eucalyptus urnigera*)** This tree is easily distinguished by its smaller stature, dark, glossy green, waxy leaves, and urn-shaped seed capsules.

PROFILE

↕ Up to 80 ft (25 m)

↔ Up to 25 ft (8 m)

🌢 Evergreen

▨ Alternate simple

◉ Tasmania

leaf up to 2 in (5 cm)
long, ¾ in (2 cm) wide

fruit up to ½ in
(1 cm) long

creamy white flower
with yellow stamens

Glaucous leaves
Both young and mature leaves
are thick, fleshy, and blue-green
with a waxy coating. They are
borne on blue-green shoots.

FEATURES

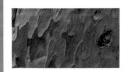

Bark Smooth, mottled gray-
white; peels in vertical strips
to reveal creamy white layer

Eucalyptus coccifera (Myrtaceae)

MOUNT WELLINGTON PEPPERMINT

Also known as the Tasmanian snow gum, this large
tree is one of the hardiest and most attractive of the
eucalyptus trees. In Tasmania it grows in the central
mountain ranges up to 4,590 ft (1,400 m) above sea level.
A broadly spreading, fast-growing species, it was first
introduced into Britain around 1840, and has been
widely planted across Europe since then.

Juvenile leaves are rounded to oval-shaped, while
mature leaves are lanceolate. The foliage emits a
peppermint fragrance when crushed. Flowers, with
numerous yellow stamens, are borne in clusters in the
leaf axils in early summer. The fruit is a small, woody,
funnel-shaped capsule that contains several seeds.
Blue-green and bloomy at first, it ripens to gray-brown.

◔ **Tasmanian blue gum (*Eucalyptus globulus*)** Unlike its
cousin, this tree has leathery, lanceolate leaves, up to 12 in (30 cm)
long and 2 in (5 cm) wide. They emit a eucalyptus scent.

PROFILE

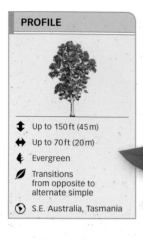

- ↕ Up to 150 ft (45 m)
- ↔ Up to 70 ft (20 m)
- ✦ Evergreen
- ⬚ Transitions from opposite to alternate simple
- ◔ S.E. Australia, Tasmania

light greenish blue young leaf with copper tint

lanceolate, mature leaf

Mountain gum leaves
Young leaves are rounded, stalkless, and borne opposite on blue-green shoots; they become lanceolate with age.

FEATURES

Bark Patchwork of gray-brown, salmon-pink, or red-brown

adult leaf up to 7 in (18 cm) long

Leaves
Lanceolate, glaucous-blue in maturity

Eucalyptus dalrympleana (Myrtaceae)

MOUNTAIN GUM

This broadly columnar tree, also known as white mountain gum or broad-leaved kindling bark, grows up to 150 ft (45 m) in its native habitat. But high in the mountains of New South Wales, 4,265 ft (1,300 m) above sea level, it seldom reaches more than half this size, and may even be shrubby.

The tree has a straight trunk with beautiful patchwork bark, which peels in long, ragged strips to reveal patches of white or fawn beneath. Juvenile and mature leaves, which vary in shape and arrangement, are often borne together on the tree. The flowers are fragrant and white, with yellow stamens. They are borne in groups of three, in umbrella-shaped flower clusters in the leaf axils in summer. The flowers are a good source of honey. Seeds are contained in a green hemispherical capsule, ripening to woody brown.

◌ **Silver gum (*Eucalyptus cordata*)** This small tree is up to 50 ft (15 m) tall. It has angular, almost square, shoots; heart-shaped juvenile leaves; and white bark mottled green and purple.

PROFILE

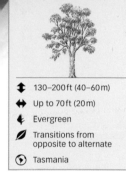

↕ 130–200 ft (40–60 m)

↔ Up to 70 ft (20 m)

🗲 Evergreen

▱ Transitions from opposite to alternate

◉ Tasmania

urn-shaped fruit, up to ½ in (1 cm) long

mature leaf up to 4 in (10 cm) long

Yellow gum leaves
Mature leaves are lanceolate, bronze when young, turning glossy apple-green with age. The fruit is a stalkless capsule, ripening from blue-green to woody brown.

FEATURES

Bark Brown or pink-red in maturity; flakes in long, linear strips to reveal paler bark

Eucalyptus johnstonii (Myrtaceae)

YELLOW GUM

This narrowly columnar tree, also known as the Tasmanian yellow gum, is one of the largest *Eucalyptus* species. It reaches up to 200 ft (60 m) in height in its native habitat. It is fast-growing, with some specimens having been known to put on more than 6 ft (2 m) of growth in a year. The tree was introduced into Europe in 1886.

The bark is blue-gray when young, and, when shed, accumulates in piles around the base of the tree's straight trunk. Juvenile leaves are rounded, thick and fleshy, and stalkless. They are borne in opposite pairs on distinctive deep red, young shoots. Mature leaves are lanceolate, bright green, and borne alternately on the stem. The flowers are fragrant, creamy white with yellow stamens, borne in umbels (clusters) of three in the leaf axils in summer. The fruit is a capsule, borne in threes. Each capsule contains several seeds.

♤ **Snow gum (*Eucalyptus pauciflora*)** This tree is distinguished from the yellow gum by its sickle-shaped leaves, up to 8 in (20 cm) long, and light gray, almost white bark.

sharply toothed
leaf margin

drawn-out,
pointed leaf tip

heart-shaped
leaf base

leaf paler
green with
hairs on
underside

PROFILE

↕ Up to 80 ft (25 m)

↔ Up to 50 ft (15 m)

🍂 Deciduous

🌿 Alternate simple

◉ China

lax, white bract, up
to 8 in (20 cm) long

Conspicuous flower bracts
Two unequal bracts enclose small
flowers with lilac anthers, which
are clustered in round heads. The
leaves are broadly ovate.

FEATURES

Bark Brown, smooth at first;
becomes fissured in maturity

circular
husk

Fruit Green,
ripening to
purple-brown

Davidia involucrata (Cornaceae)

POCKET HANDKERCHIEF TREE

This Chinese tree was introduced in the West by
English plant collector Ernest H. Wilson in 1901. It
is also known as the dove tree or ghost tree. All three
common names refer to the hanging white bracts, which
surround the flower in spring.

The brown bark, initially
smooth, develops orange-brown
cracks, and fissures into irregular
plates. Shiny, bright green leaves
grow up to 6 in (15 cm) long and
curl in hot or dry weather to
reduce water loss. The flower
heads are up to ½ in (1 cm) wide.
The fruit is a green, circular husk,
borne singly on a 1½ in- (4 cm-)
long stalk. It contains a hard nut
with several seeds inside.

Showy blossom
In spring, the flower bracts
are conspicuous on this
large, broadly conical,
sometimes spreading tree.

flower up to
½ in (1 cm) long

oblong to ovate leaflet,
up to 4 in (10 cm) long

heavily lobed
leaflet has
toothed margin

Foliage and flowers
Up to 18 in (45 cm) long, the
pinnate, and sometimes bipinnate,
leaf carries 11 to 13 leaflets. The
flowers appears in summer.

PROFILE

- Up to 40 ft (12 m)
- 40 ft (12 m)
- Deciduous
- Alternate compound
- China, Korea

FEATURES

young
seed
capsule

Fruit Bladderlike seed
capsule, green ripening to
yellow-brown

Koelreuteria paniculata (Sapindaceae)

PRIDE OF INDIA

Also known as the golden-rain tree or China tree,
this medium-sized, broadly spreading tree is native to
China and Korea. It has, however, been widely planted
in parks, gardens, and arboreta throughout India and
the temperate world since its introduction into wider
cultivation in 1763.

The young bark is pale brown
and smooth, becoming fissured
in maturity. Pinnate leaves are
matte and dark green on the
upper surface, paler with a pink
midrib on the underside. They are
borne in spring on long and
arching shoots, which gradually
fade from red through pink to
light green by high summer.
Clusters of bright golden yellow
flowers are carried in open
panicles, up to 18 in (45 cm) long.

Ornamental foliage
Planted for its flowers and
unusual fruit, the pride of
India tree has a graceful,
broadly spreading form.

PROFILE

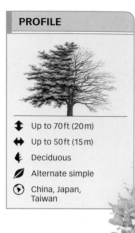

⬍ Up to 70 ft (20 m)

⬌ Up to 50 ft (15 m)

🍂 Deciduous

▰ Alternate simple

◉ China, Japan, Taiwan

Flower cluster
In early summer, tiny, four-petaled flowers are borne in clusters (corymbs) above the foliage, at regular intervals along horizontal branches.

oval leaf, up to 6 in (15 cm) long, with dark green, shiny upper surface

small, creamy white flower

erect flower stalk

hairy, pale blue-green leaf underside

FEATURES

flower buds in cluster

Flower buds Green, in broad, flattened corymbs, up to 15 cm (6 in) wide

Fruit Rounded, blue-black berries, on pink-red stalks

Cornus controversa (Cornaceae)

WEDDING-CAKE TREE

Also known as the table dogwood, this broadly spreading, medium-sized tree came into western cultivation before 1880. Its common names refer to its symmetrical, horizontal branches, which gradually grow shorter toward the top of the tree in tiers, like the layers of a wedding cake.

It is one of the only two dogwoods to have alternate leaves, while the rest have opposite leaves. The bark is smooth, gray to light brown, and becomes lightly fissured in maturity. The distinctly veined leaves turn rich plum-red in fall before being shed. The flowers are followed by clusters of small berries in fall.

◈ **Variegated table dogwood**
This cultivar (*Cornus controversa* 'Variegata') has narrower, variegated leaves with a striking silver-cream margin.

PROFILE

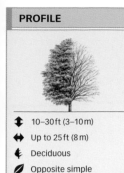

- ↕ 10–30 ft (3–10 m)
- ↔ Up to 25 ft (8 m)
- 🍂 Deciduous
- 🌿 Opposite simple
- ⊙ Japan, China, Korea

tiny, green flower bobble

bract, up to 2 in (5 cm) long, tapers to a point

untoothed leaf margin

roughly oval leaf, up to 3 in (7.5 cm) long

Bract-enclosed flowers
Flower clusters are surrounded by four creamy white or pink-white bracts, positioned on upright stalks above the dark green foliage.

FEATURES

Leaves Pink, red, and purple in fall before being shed

Fruit Red, globular; borne in clusters on long stalks

Cornus kousa (Cornaceae)

JAPANESE DOGWOOD

Also known as the Japanese strawberry tree, this tree is commonly planted in gardens throughout the temperate regions of the world. Indigenous to Japan, it was introduced into western cultivation in 1875.

A Chinese variety of this tree was discovered by Ernest H. Wilson in 1907, and is also cultivated in Europe.

The Japanese dogwood has a rich red-brown bark, which flakes to reveal cream- or fawn-colored new bark beneath. The leaves have a shiny upper surface and a paler underside, with tufts of brown hairs in the leaf vein axils. The tiny, green flowers appear in early summer, and develop into strawberry-like, edible fruit in late summer.

Chinese dogwood
This large tree (*Cornus kousa* var. *chinensis*) has larger leaves, which are smooth around the margin, and creamy white flower bracts, which turn pink.

PROFILE

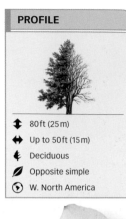

- ↕ 80ft (25m)
- ↔ Up to 50ft (15m)
- 🌿 Deciduous
- ⬭ Opposite simple
- ◉ W. North America

Erect flowers
Flowers are borne at shoot ends on erect stalks, enclosed by four to seven bracts. The variable leaves are obovate to oval in shape.

creamy white, often blushed pink, petal-like bract, up to 3in (7.5cm) long

dark green leaf, up to 6in (15cm) long, 3in (7.5cm) wide

cluster of small, green-purple flowers with yellow stamens

elliptic, pointed leaf with untoothed margin

FEATURES

Leaves Turn yellow or red in fall before being shed

Fruit Rounded, strawberry-like, carried on long stalks

Cornus nuttallii (Cornaceae)

PACIFIC DOGWOOD

With a natural habitat ranging from British Columbia to California, the Pacific dogwood has been cultivated in parks and gardens outside this region since its discovery in 1835. In cultivation, this tree rarely attains its full height, often growing as a large, multi-stemmed shrub, or small tree up to 30ft (10m) tall. This medium-sized, broadly conical tree produces the largest bracts of any dogwood species.

The smooth purple-gray bark cracks into irregular plates in maturity. The leaves are pale green, with a fine coating of hairs on the underside. In late spring, flowers appear in dense hemispherical clusters, up to 1in (2.5cm) wide, followed by small, red fruit.

⚘American flowering dogwood
This smaller related tree (*Cornus florida*) has four bracts that enclose flower heads in winter.

PROFILE

‡ Up to 50 ft (15 m)

↔ Up to 50 ft (15 m)

🍂 Deciduous

🌿 Opposite compound

◐ N. and N.E. China, Korea, Japan

glossy green upper surface of leaf

leaflet 4 in (10 cm) long

yellow-green leaf stalk

midrib of leaf tinged yellow

Leaf and leaflets
The pinnate leaves of the Amur cork tree grow up to 16 in (40 cm) long. Each leaf consists of up to 13 ovate leaflets.

FEATURES

Bark Gray, thick, deeply fissured, and corky

berry up to ⅜ in (0.8 cm) wide

Fruit Stalked clusters of smooth, rounded, green berries, ripen to shiny black

Phellodendron amurense (Rutaceae)

AMUR CORK TREE

As implied by its name, the bark of this tree has been used in China in much the same way as that of the cork oak (p.170). It is native to Amur, a region straddling the border between China and Russia, as well as parts of Korea and Japan. The Amur cork tree has been in western cultivation since 1885. However, in recent years it has become an invasive species, and is now listed as a noxious weed in some states in the US, such as Massachusetts.

Its pinnate leaves may look like those of the common ash (p.300) with leaflets up to 2 in (5 cm) wide. The underside of each leaflet is pale green, with tufts of hair at the base of the midrib. The flowers are small and white, with yellow anthers, borne on terminal panicles up to 6 in (15 cm) long.

Spreading form
Until recently, the broadly spreading, medium-sized Amur cork tree was often planted in the United States.

Flowers in corymbs
Pungently scented flowers are carried in domed corymbs (flat-topped flower clusters) up to 6 in (15 cm) wide.

dark green upper surface of leaflet

leaflet up to 5 in (12 cm) long

creamy white flowers, with yellow anthers

leaf stalk fades from deep pink to yellow-green

pale green underside of leaflet

PROFILE

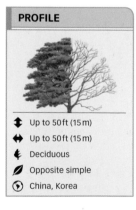

⬍ Up to 50 ft (15 m)

⬌ Up to 50 ft (15 m)

🗲 Deciduous

🗌 Opposite simple

⊙ China, Korea

Tetradium daniellii (Rutaceae)

EUODIA

This tree is also called the bee-bee tree because its flowers attract honeybees and appear at a time when few other trees are in bloom. Its botanical name commemorates William Daniell, a British army surgeon and botanist who, in the 1860s, collected specimens of the tree in China. However, it was not introduced into western cultivation until 1905.

The bark is gray and smooth, even in maturity. Pinnate leaves, up to 16 in (40 cm) long, carry 7 to 11 elliptic, oval to lanceolate leaflets. The leaves turn pale yellow in fall before being shed. In summer, small flowers are followed by clusters of spherical fruit, which ripen from red-brown to black. Each fruit contains tiny, black seeds.

Fast-growing tree
The euodia is a broadly spreading, medium-sized, deciduous tree. It grows rapidly to become as wide as it is tall.

PROFILE

⬍ Up to 70 ft (20 m)

⬌ Up to 50 ft (15 m)

🍃 Evergreen

🌿 Alternate simple

⊙ Europe, W. Asia, N. Africa

glossy red berry, up to ½ in (1 cm) wide

leaf up to 4 in (10 cm) long

Densely clustered berries
Fruit of the common holly tree
is a spherical, smooth berry, borne
in dense groups clustered along
the shoots in winter.

FEATURES

male flowers

Flowers Purple-tinged or
white; male and female appear
on separate trees

Ilex aquifolium (Aquifoliaceae)

COMMON HOLLY

This easily recognizable, broadly columnar,
medium-sized tree is extremely hardy. Its dense foliage
provides welcome shelter in exposed mountainous
and coastal locations. Holly has long been cultivated in
gardens and has given rise to numerous, common garden
cultivars. In the western world, holly is closely linked
to Christmas celebrations.

The bark is silver-gray and smooth even in maturity.
Elliptic to ovate leaves are thick and stiff. They are glossy,
dark green, and waxy on the upper surface; light green
on the underside; and variable around the margin. Some
leaves bear sharp, strong spines, others may have no
spines at all. Male and female flowers are borne on
separate trees. They are small, clustered into the leaf
axils, and appear in early summer. The fruit ripen from
green to bright red and are eaten by birds in winter.

◇ **Horned holly (*Ilex cornuta*)** It is easy to distinguish this
Chinese holly from the common holly by its distinct, rectangular
leaves displaying up to five angular spines.

leaf up to 1½ in (4 cm) long

sharply pointed, central lobe of leaf

Angular lobes
The leaves have three to seven angular lobes, all ferociously spined, with the central lobe elongated and pointed.

FEATURES

Bark Slate-gray, smooth

Fruit Bright red, spherical berries; carried in clusters in fall

Ilex pernyi (Aquifoliaceae)

PERNY'S HOLLY

This narrowly conical tree, sometimes large shrub, is native to central and western China. It was discovered in 1858 by Abbé Paul Perny, a French Jesuit missionary. However, it was another 40 years before it was introduced into western cultivation. Today, it is commonly planted in botanical gardens and arboreta.

The leaves, which are borne on thick, leathery stalks, are deep glossy green on the upper surface and paler on the underside. They are oblong, sometimes almost triangular, and occasionally trident-shaped. The branches of Perny's holly are erect, ascending, and then arching. The flowers are cream to pale yellow and slightly fragrant. Male and female flowers are borne on separate trees, in tight clusters, on the leaf axils. The fruit is a rounded, red berry, carried on erect shoots. Borne in fall, it persists on the tree well into winter.

♧ *Ilex corallina* This tree is easily distinguished from Perny's holly by its wavy-edged, lanceolate leaves, up to 3 in (8 cm) long. These are normally spineless, though toothed.

PROFILE

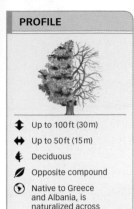

- Up to 100 ft (30 m)
- Up to 50 ft (15 m)
- Deciduous
- Opposite compound
- Native to Greece and Albania, is naturalized across temperate regions

creamy white flower with five petals

flowers in branched cluster

yellow blotch on flower, turns pink

Conical panicles
The creamy white flowers are borne in large clusters in spring. These are conical and upright, up to 10 in (25 cm) tall.

FEATURES

Leaves
Compound palmate, bright green, strongly veined

Fruit Rounded green husk with spines; up to 2½ in (6 cm) wide

Aesculus hippocastanum (Sapindaceae)

COMMON HORSE CHESTNUT

Perhaps one of the most common large trees in cultivation, the common horse chestnut is indigenous only to Greece and Albania. This broadly columnar tree has been widely planted in the temperate world since its introduction in 1650. In recent years, it has come under attack from a leaf-mining moth and a bleeding canker bacterium, both of which are causing widespread damage to trees across Europe.

The bark is gray to brown-orange and smooth, turning shallowly fissured and scaly in maturity. Leaves consisting of up to seven obovate leaflets, each up to 10 in (25 cm) long, turn golden brown before being shed in mid-fall. Winter buds are large, rich red-brown, and sticky. The rounded fruit contains one to three seeds.

◊ **California buckeye (*Aesculus californica*)** This American species has much smaller leaves and flowers than the common horse chestnut, and its flowers are borne in summer.

Compound palmate leaves
The leaves have five to seven leaflets, all meeting the leaf stalk at a common point.

rough, slightly pear-shaped, green fruit

pink petiole (leaf stalk)

obovate to broadly lanceolate leaflet, up to 10 in (25 cm) long

PROFILE

↕ 80–100 ft (25–30 m)
↔ Up to 40 ft (12 m)
🍂 Deciduous
🌿 Opposite compound
🌍 N. India, N.W. Himalayas

FEATURES

Flowers White to pale pink; borne on erect panicles up to 10 in (25 cm) long

brown seed

Seed Dark brown, enclosed in green husk

Aesculus indica (Sapindaceae)

INDIAN HORSE CHESTNUT

This large, broadly columnar tree is native to north India and the northwest Himalayas, where it grows in the foothills up to 9,840 ft (3,000 m) above sea level. Introduced into western cultivation in 1851, it is planted in parks, gardens, and arboreta across the temperate world, especially Europe and North America.

The bark is gray-brown and smooth, even in maturity. Bronze-colored leaflets emerge in spring, turning glossy grass green in summer and yellow in fall. Borne in summer, the flowers have a red or yellow blotch at the base. They are more slender than the flowers of the common horse chestnut (p.281). The fruit is a green husk, ripening to brown. It opens to reveal up to three seeds, smaller than those of the common horse chestnut.

♀ **Chinese horse chestnut (*Aesculus chinensis*)** This tree is smaller than the Indian horse chestnut, with smaller leaves and white flowers that are borne on thinner panicles.

PROFILE

red blotch on
flower base

Flowers and leaves
Creamy white flowers, amidst
shiny dark gray-green leaves,
are borne in spring.

flowers borne in
erect panicle, up
to 10in (25cm) long

↕ 70–100ft (20–30m)

↔ Up to 50ft (15m)

🍂 Deciduous

🌿 Opposite compound

📍 Japan

FEATURES

Fruit Almost spineless, up to
½in (1.5cm) long

Aesculus turbinata (Sapindaceae)

JAPANESE HORSE CHESTNUT

This broadly columnar tree is similar in appearance to
the common horse chestnut (p.281). It is widely planted
in its native Japan as an ornamental species in parks
and gardens. Elsewhere, it is not as common although
regularly found in botanic gardens and arboreta across
temperate regions of the world. It is slower-growing than
the common horse chestnut, with smaller flowers, which
are not carried in such profusion. However, it has larger
leaves, which turn bright orange in fall.

The bark is brown and smooth when young, flaking in
maturity. Winter buds are red-brown, glossy, and sticky.
The ovate leaves are compound palmate, consisting of
a maximum of seven deeply veined leaflets up to 16in
(40cm) long. They have fine, uneven toothed margins and
rusty hairs along the midrib on the pale green underside.
The egg-shaped fruit ripens to brown and cracks to
reveal up to three shiny dark brown seeds.

Conical panicles
Flowers ranging in color from pink to deep crimson are borne on branched, conical flower clusters.

flower with five petals

flowers borne on panicle up to 8 in (20 cm) long

PROFILE

Up to 70 ft (20 m)

Up to 30 ft (10 m)

Deciduous

Opposite compound

Germany

FEATURES

Leaves Compound palmate, made of five to seven leaflets

fruit up to 2 in (5 cm) wide

Fruit Rounded, green husk; ripening to dull brown

Aesculus x *carnea* (Sapindaceae)

RED HORSE CHESTNUT

With its bright pink-red flowers, this popular hybrid tree is easily told apart from the common horse chestnut (p.281), which is one of its parents, the other being the American red buckeye (p.286). Little is known about the origins of this broadly columnar species, but it possibly occurred naturally in Germany in the early 1800s.

The bark is dull brown and smooth when young, developing shallow fissures and breaking into flaking plates with age. The mid- to dark green leaves consist of obovate leaflets up to 10 in (25 cm) long. They have pronounced parallel veins and are joined at the base to a long, pink-green stalk. The fruit, sometimes covered in spines, splits in segments to reveal up to three shiny brown seeds.

Blooming in spring
The canopy of the large red horse chestnut starts blooming from mid- to late spring.

Yellow flower clusters
In early summer, pale yellow flowers are borne on stalks in erect, elongated, panicles (branched clusters) that are up to 6 in (15 cm) long.

pink blotch on flower

FEATURES

shallowly serrated leaflet margin

Leaves Orange-red in fall before being shed

Fruit Green ripening to brown; carry up to three seeds

Aesculus flava (Sapindaceae)

SWEET BUCKEYE

This large, round-headed but broadly conical tree is native to the southeastern United States, from where it was introduced into Europe in 1764. It is the only buckeye (or horse chestnut) to produce yellow flowers and is sometimes referred to as the yellow buckeye.

The bark is brown-gray, flaking in maturity into large, irregular-shaped scales. Branches tend to be horizontal, sometimes drooping, with a characteristic upward sweep at the tip. The leaves are compound palmate, each consisting of five bright to deep green leaflets, which are narrowly elliptic and up to 6 in (15 cm) long. They are joined to a pea-green, sometimes pink, leaf stalk at a common point. The fruit is a smooth, rounded to oblong husk, up to 2 in (5 cm) long. It ripens to split in segments to reveal two to three shiny brown seeds.

◇ **Ohio buckeye (*Aesculus glabra*)** This tree can be distinguished from the sweet buckeye by its tall, thin, pale green-yellow flowers and its elliptic leaflets, which are up to 8 in (20 cm) long.

Erect flowers
The red buckeye has four-petaled flowers, which bloom in early summer. The leaves are compound palmate with elliptic leaflets.

slender, bright crimson-red flower

flowers borne in erect, open panicle, up to 8 in (20 cm) long

leaflet up to 6 in (15 cm) long

⬍ Up to 22 ft (7 m)
⬌ Up to 10 ft (3 m)
❧ Deciduous
∅ Opposite compound
◉ S.E. United States

FEATURES

leaf tapers to a point

Leaves Five glossy, dark green leaflets, each on a short stalk

fruit up to 2 in (5 cm) long

Fruit Pear-shaped, russet-green ripening to light brown; appears after flowers

Aesculus pavia (Sapindaceae)

RED BUCKEYE

Native to the coastal plains of the southeastern United States, from North Carolina to Florida, and west to the Mississippi River basin, this tree grows in moist woods and thickets. Sometimes no more than a shrub, it is a small, broadly spreading tree. The red buckeye tree is one of the parents of the hybrid red horse chestnut (p.284) and gives the red coloring to the latter's flowers. It was introduced into the western world in 1711 and has been cultivated in parks, gardens, and arboreta since then.

The red buckeye has dark gray bark, smooth at first, which cracks into irregular plates in maturity. The leaves are made up of sharply toothed, elliptic leaflets. Each leaflet is borne on a yellow-pink petiole (leaf stalk) up to 7 in (18 cm) long. In fall, the leaves turn orange and red. This tree produces small fruit, which open to reveal one or two glossy, mahogany-colored seeds.

⚪ **Sunrise horse chestnut (*Aesculus* x *neglecta* 'Erythroblastos')** This tree is distinguished by its salmon-pink leaves in spring and panicles of creamy yellow flowers.

leaf dark green on upper surface

Leaves and fruit
The palmate, lobed leaves of the field maple have yellow-green veins on the upper surface. The fruit is borne in pendulous clusters.

↕ Up to 70 ft (20 m)

↔ Up to 30 ft (10 m)

🌢 Deciduous

🌿 Opposite simple

⊙ Europe, N. Africa, S.W. Asia

two winged seedcases (keys) arranged in pairs

leaf borne on green or pink-green petiole

leaf up to 3 in (7.5 cm) long and wide

FEATURES

Bark Pale gray-brown, smooth; becomes corky in maturity

young leaf

Flowers Small, greenish yellow; borne in erect terminal clusters in spring

Acer campestre (Sapindaceae)

FIELD MAPLE

This medium-sized, broadly columnar tree is normally shorter than its maximum height of 70 ft (20 m). Also known as hedge maple, this tree, along with common hawthorn (p.243), serves as a field boundary hedge across its natural range. Some European specimens of this long-lived tree are known to be over 500 years old.

The bark becomes corky with age and this extends to relatively small branches, giving them a "winged" appearance. The palmate leaves have five rounded lobes and are pale green on the underside. The leaf stalk, or petiole, exudes a milky liquid when crushed or broken. In fall, the leaves may turn bright golden yellow or orange before falling. The tree produces a typical green, two-winged seedcase, which ripens to light brown, sometimes remaining on the tree into early winter.

♀ **Rough-barked maple (*Acer triflorum*)** This related species is distinguished from the field maple by its trifoliate leaves and heavily fissured bark.

PROFILE

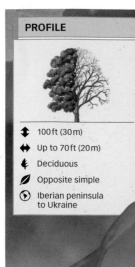

- 100 ft (30 m)
- Up to 70 ft (20 m)
- Deciduous
- Opposite simple
- Iberian peninsula to Ukraine

coarsely toothed leaf margin

Leaves and flowers
Sycamore leaves are borne on pink-green petioles. The flowers hang in dense, pendulous clusters, appearing at the same time as leaves in spring.

flower cluster up to 3 in (7.5 cm) long

FEATURES

leaf up to 8 in (20 cm) wide

Leaves Palmate with five lobes; deep green upper surface, pale green underside

wing 1 in (2.5 cm) long

Fruit Two-winged seed, borne in clusters in early summer

Acer pseudoplatanus (Sapindaceae)

SYCAMORE

In some parts of the world this large, broadly columnar tree is known as the plane tree, although it is not part of the *Platanus* genus. The sycamore is native to a region extending from the Iberian Peninsula to the Ukraine. Although not native to Britain, it is now extremely common there, having become naturalized many centuries ago. A hardy tree, it is resistant to strong winds and exposure to salt-laden air in coastal regions.

Its gray bark is smooth at first, becoming grayish-pink in maturity with irregular flaking plates. The leaves turn yellow in fall. Lime-green leaf buds are very conspicuous in winter. The fruit, a two-winged seed, is pink-green when unripe and matures to light brown in fall.

☘ *Acer pseudoplatanus* **'Brilliantissimum'**
Unlike the sycamore, this small, slow-growing cultivar produces dramatic shrimp-pink leaves in spring

PROFILE

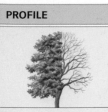

⬍ Up to 100 ft (30 m)

⬌ Up to 70 ft (20 m)

🍂 Deciduous

🍃 Opposite simple

⊙ Europe, S.W. Asia

bright green leaf upper surface

leaf ends in slender, sharp tip

sharply toothed leaf margin

Foliage and seeds
The palmate leaves with five angular lobes are borne on long, slender, pink-yellow petioles. The seeds develop in early fall.

two-winged fruit

FEATURES

Bark Light fawn-gray and smooth when young; develops ridges and fissures in maturity

Flowers Small, bright yellow, sometimes red; borne in spring as leaves emerge

Acer platanoides (Sapindaceae)

NORWAY MAPLE

This fast-growing tree is native to southwestern Asia and Europe, including southern Scandinavia. It has been widely cultivated within its native region and elsewhere in the temperate world for centuries. Broadly conical, it is planted as an ornamental tree in parks, gardens, urban areas, and alongside roads.

The Norway maple is prone to damage from gray squirrels, which strip the bark, sometimes killing large branches across the crown. The leaves, glaucous on the underside, are up to 6 in (15 cm) long and wide. In fall they turn striking shades of gold, orange, and red before falling. Small flowers are borne in conspicuous drooping clusters in spring. The green-yellow, often pink-blushed, fruit is a typical *Acer* two-winged seed. It is borne in pendulous clusters, up to 2 in (5 cm) long, ripening to a fawn color in fall.

♧ *Acer platanoides* **'Schwedleri'** This cultivar is distinguished from the Norway maple by its rich crimson-purple-colored leaves, which are especially vibrant in spring.

Palmate leaves
The leaves have five to seven broad lobes, each running to a fine point. The fruit—a two-winged seed—is borne in clusters.

lobe ends in fine point

leaf up to 4 in (10 cm) long

two-winged seed

FEATURES

bronze leaf tip

Leaves Butter-yellow in fall before being shed

erect flower cluster

Flowers Small, greenish yellow, appear in spring

Acer cappadocicum (Sapindaceae)

CAUCASIAN MAPLE

An attractive large maple, this tree is native to Turkey, Iran, the Caucasus between Europe and Asia, and western China. The large, stately, broadly spreading Caucasian maple was introduced into western Europe in 1838. Since then, it has become a popular choice for planting in parks, gardens, and arboreta.

The bark is similar to the common ash (p.300), light gray and smooth at first, developing shallow, vertical fissures in maturity. The leaves are palmate, growing up to 6 in (15 cm) wide. Grass-green on the upper surface, they are paler on the underside, with tufts of hairs in the vein axils. The fruit is green when young, growing up to 1½ in (4 cm) long. It turns light brown on ripening and is shed from the tree in late fall.

Top-heavy tree
The Caucasian maple has a rounded crown borne on a little-branched trunk, about 12 ft (4 m) from ground level

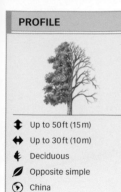

- Up to 50 ft (15 m)
- Up to 30 ft (10 m)
- Deciduous
- Opposite simple
- China

Variable leaves
The faintly lobed leaves are variable in shape, but are often rounded at the base. The fruit is a two-winged seed.

leaf up to 6 in (15 cm) long, 3 in (8 cm) wide

wing of seedcase angled downward

irregularly serrated leaf margin

leaf ends in short, curved point

FEATURES

Bark Smooth, with vertical white and green stripes

Leaves Turn orange-red, before being shed in fall

Flowers Yellow-green, borne in clusters

Acer davidii (Sapindaceae)

PÈRE DAVID'S MAPLE

Native to China, this tree is often referred to as Snake-bark maple—a generic term for a whole group of the *Acer* genus. From China, it was introduced into the West in 1879 by English botanist Charles Maries. A small, broadly columnar tree, it was named after French Jesuit missionary and plant collector Père (Father) David. Many clones and cultivars, such as 'George Forrest' and 'Ernest Wilson,' are now grown in parks and gardens.

The variable leaves range from ovate to oblong-lanceolate. Small flowers emerge in spring just after the leaves. Green-yellow, often blushed pink fruit ripen to light-brown. They are borne in pendulous clusters up to 3 in (7.5 cm) long in fall.

♧ **Père David's maple 'George Forrest'**
Spreading branches and dark green leaves attached to shoots by red petioles distinguish this cultivar (*A. Davidii* 'George Forrest').

PROFILE

- 50 ft (15 m)
- Up to 20 ft (6 m)
- Deciduous
- Opposite compound
- C. China

Fruit and leaves
This tree produces pairs of pale green, winged seeds, covered with soft down. Its leaves have three elliptic leaflets.

blunt teeth on each side of leaf

blue-green leaf underside

leaf dark green on upper surface

leaf up to 4 in (10 cm) long

wing approximately 1¼ in (3 cm) long, ripens to light brown

FEATURES

Bark Cinnamon-red; smooth at first, peels in papery flakes

closed, small, green-yellow flower

Flowers Borne on hairy stalks, in drooping clusters in spring

Acer griseum (Sapindaceae)

PAPERBARK MAPLE

Probably one of the finest trees for bark color, the small, broadly columnar paperbark maple is native to central China. It was discovered by Ernest H. Wilson in 1901, and almost immediately became a garden favorite throughout the temperate world.

Its bark is distinctive in winter, when low sunlight shines through the translucent bark peelings, which appear in thin flakes and curl back on themselves. The leaves are glaucous blue-green on the underside, with a light covering of hairs, reminiscent of oaks in shape. In fall, they turn from orange, through red, to burgundy before falling. The fruit ripens from green to light brown, remaining on the tree well into winter.

✿ Golden Japanese maple
Acer shirasawanum 'Aureum' has a plain, light brown bark and golden, palmate leaves.

PROFILE

- Up to 100 ft (30 m)
- Up to 50 ft (15 m)
- Deciduous
- Alternate simple
- S.E. Canada, N.E. United States

leaf up to 4 in (10 cm) long and wide

Fall colors
In fall, dark matte green, palmate leaves of the red maple turn a vibrant, bright scarlet red before falling.

shallow-lobed leaf

lobe reaches halfway to base of leaf

coarsely serrated leaf margin

FEATURES

Flowers Small, bright red; borne in clusters on short stalks in early spring

Acer rubrum (Sapindaceae)

RED MAPLE

The large, broadly columnar red maple has a natural range from eastern Newfoundland and Ontario in Canada to Florida and Texas in the United States. This tree is normally found growing in deep, damp soils.

The bark remains dark gray and smooth in maturity. The leaves have three or five lobes with glaucous blue-green, almost white, undersides, and yellow hairs around the leaf veins. The showy flowers are borne on bare branches before the leaves appear. These are followed by pairs of pale green, downy, winged seeds, each wing being around 1 in (2.5 cm) long. These turn red and green, ripening to light brown, and stay on the tree into winter.

Vibrant fall hue
The striking red maple is one of the main constituents of American fall leaf color displays.

pair of winged seeds, up to 1cm (½in) long

red-tinged fruit wing

leaf margin has forward-facing serration

palmate leaf, up to 4in (10cm) wide, with pointed lobes

Japanese maple foliage
Slender shoots end in paired, lobed leaves varying in shape. Clusters of rich green, winged fruit ripen to red in summer.

FEATURES

variable fall color

Leaves Five or seven pointed lobes; bright red, orange, or yellow in fall

Acer palmatum (Sapindaceae)

JAPANESE MAPLE

The smooth Japanese maple was first known to Europeans in 1783. However, a trade embargo with Japan delayed its introduction into western cultivation until 1820. This tree is widely planted throughout the world and has given rise to hundreds of garden cultivars.

It has a smooth, gray-brown bark, even in maturity. Some variants have leaves that are cut in between the lobes almost to the midrib. In spring, as new leaves emerge, small burgundy-red flowers with yellow stamens are borne in upright, stalked clusters, becoming drooping later. The winged fruit are borne in clusters of up to 20 seeds. Some remain on the tree even after the leaves have fallen.

◑ **Japanese maple cultivar**
This cultivar (*A. palmatum* 'Osakazuki') has bright red fall foliage and regular, seven-lobed leaves.

Fall leaves
Palmate, almost rounded, leaves have 5 to 11 pointed lobes. Green at first, leaves turn to red, orange, yellow, or wine-purple in fall.

toothed lobe tapers to a point

FEATURES

Flowers Purple-red, with lemon-yellow stamens; borne in pendulous, stalked clusters

Seeds Winged; borne in pairs on long stalks

Acer japonicum (Sapindaceae)

FULLMOON MAPLE

Native to the Japanese islands of Hokkaido and Honshu, this tree was introduced into western cultivation in 1864. Today it is rarely seen in cultivation, having been replaced by many of its own cultivars such as 'Aconitifolium' and 'Vitifolium.' It has a sprawling form, with low branches to ground level.

The bark is silver gray-brown and smooth, even in maturity. The slightly crinkly leaves have a pale green, hairy underside. Its toothed leaves are up to 5 in (12.5 cm) long and wide. Small flowers are borne in hairy clusters just as the leaves emerge in spring. Green winged seeds, up to 1 in (2.5 cm) long, ripen to red and finally brown in late summer.

◊ **Vine-leaved Japanese maple**
This tree (*Acer japonicum* 'Vitifolium') has larger vine-shaped leaves with up to nine shallowly cut lobes.

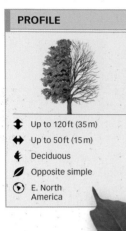
palmate leaf, up to 6 in (15 cm) long and wide

Five-lobed leaves
Sugar maple leaves have five prominent, tapered lobes that run to long points. The leaves have a mid- to dark green upper surface.

leaf lobe tapers to a point

FEATURES

Bark Smooth, brown-gray; becomes shallowly fissured, flaking in maturity

Leaves Bright scarlet, gold, or yellow in fall, before being shed

Acer saccharum (Sapindaceae)

SUGAR MAPLE

This large, broadly columnar tree is native to eastern North America, from Quebec south to Missouri. Besides being known for maple syrup, the sugar maple is also known for its timber, which is hard and resistant to wear, and has been used to make dance floors. It is a common tree in cultivation outside its natural range, both in North America and throughout the northern hemisphere, in parks, gardens, and arboreta.

Sugar maple leaves have a pale green underside, with hairs in the leaf axils. Small, yellowish green flowers are borne in drooping, open clusters in early spring, just as the new leaves appear. The fruit is a pair of winged seeds, up to 1 in (2.5 cm) long, bright green ripening to brown and persisting on the tree well into fall.

Sugary sap
Maple syrup is made from sap tapped from the bark of the sugar maple. The sap is boiled to produce syrup.

PROFILE

Up to 100ft (30m)

Up to 50ft (15m)

Deciduous

Opposite simple

E. North America

Silver maple foliage
The leaves have five large lobes ending in a sharp point. They are held on lax petioles, or leaf stalks, up to 6in (15cm) long.

sharply toothed leaf lobe

palmate leaf

lobed leaf, up to 6in (15cm) long and wide

FEATURES

Bark Smooth, gray; becomes flaky; shoots may emerge from beneath it

yellow fall leaf

Leaves Palmate; turning bright yellow in fall

Acer saccharinum (Sapindaceae)

SILVER MAPLE

One of the fastest growing maples in North America, the silver maple is a broadly columnar, large tree with a natural range that extends across eastern North America. Since its introduction into Europe in 1725, it has been widely planted in parks and arboreta. This tree is sometimes referred to as the American version of a sycamore (p.288), but it has a more refined habit and foliage.

The broad leaves have a light green upper surface and a blue-green to silver underside with some hairs. Small, greenish yellow flowers, without petals, are clustered on young shoots as leaves emerge in spring. The fruit are winged seeds, up to 1in (2.5cm) long, bright green ripening to brown.

Elegant habit
The silver maple has a light, open crown and bicolored leaves, which catch the light as they flutter in the breeze.

PROFILE

- ↕ 50–70ft (15–20m)
- ↔ Up to 30ft (10m)
- 🍂 Deciduous
- 🌿 Opposite compound
- 🕐 C. North America

Box elder leaves
The leaves of the box elder, unlike many related maple trees, are pinnate. Each leaf consists of up to seven variably shaped leaflets.

bright green leaf upper surface

hairy, light green leaf underside

leaf up to 4in (10cm) long

FEATURES

flowers borne in tassel-like clusters

Flowers Small, yellow-green, with drooping stamens

Fruit Downward-pointing seed with two wings; up to ¾in (2cm) long

Acer negundo (Sapindaceae)

BOX ELDER

Also called the ash-leaved maple, the box elder is a broadly columnar, light, and open-branched tree. It grows wild across central North America, particularly alongside rivers and where soil is moist. The species was introduced into Europe as early as 1688.

Box elder bark is brown to silver-gray, thin and smooth when young, remaining relatively so into maturity. The leaves are similar to those of the unrelated common elder (*Sambucus nigra*). The leaflets are sometimes lobed and arranged along leaf stalks in pairs, with one larger, final leaflet at the end. Their fall color tends to be yellow and muted. Small male and female petal-less flowers are borne on separate trees in spring as the leaves emerge. They are carried in drooping clusters from the outermost twigs. The fruit is typical of the *Acer* genus—a seed with two incurved wings, arranged in pairs.

☘ **Henry's maple (*Acer henryi*)** Unlike box elder, its pinnate leaves consist of three narrower leaflets. Borne on pink petioles (stalks), they develop a bright red and orange tint in fall.

lobe cut deep into the leaf center

young green winged fruit

coarsely toothed leaf margin

Leaves and fruit
Each palmate leaf of the Oregon maple has either five or seven large lobes. The fruit has paired wings set at right angles.

flower cluster, up to 8 in (20 cm) long

Flowers Yellow-green, borne in conspicuous, pendulous clusters in spring

Acer macrophyllum (Sapindaceae)

OREGON MAPLE

Characterized by its large leaves, the Oregon maple is indigenous to a region stretching from British Columbia, Canada, to California. This majestic, large, broadly columnar tree was introduced into cultivation in 1812.

The bark is smooth and gray-brown, becoming heavily and vertically fissured in maturity. On older specimens, the trunk may grow up to 5 ft (1.5 m) wide at chest height. The leaves are up to 10 in (25 cm) long and 12 in (30 cm) wide. They are shiny dark green on the upper surface and paler on the underside. Each leaf is carried on a long, buff-colored, sometimes red, petiole (leaf stalk). The leaves turn deep yellow and orange to brown in fall. Slightly fragrant flowers emerge around the same time as the leaves begin to grow in spring. The young, green fruit has an initial covering of fawn-colored hairs. It has large, smooth wings up to 2 in (5 cm) long. It ripens to brown, and persists on the tree well into fall.

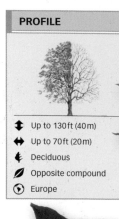

⬍ Up to 130 ft (40 m)

⬌ Up to 70 ft (20 m)

🍂 Deciduous

🌿 Opposite compound

◉ Europe

white, faintly hairy
midrib on underside
of leaflet

shallow-toothed
leaflet margin

distinctive
black leaf bud

winged seed up to
1½ in (4 cm) long

leaflet up to
5 in (12 cm) long

Ash keys
The fruit is a pendulous cluster
of flattened, winged seeds known
as keys. Initially green, it turns light
brown to gray when mature.

FEATURES

Flowers Appear in late
winter clustered along
previous year's shoots

Fraxinus excelsior (Oleaceae)

COMMON ASH

Native to most of Europe, from the Pyrenees to
the Caucasus, this broadly columnar tree is one of the
largest European deciduous trees. It is a common
species in woodland, field boundaries, and open
moorland, particularly on calcareous limestone soil. It
has become invasive in some urban areas, seeding itself
with ease into gardens and parks. It has a light, airy
crown with relatively few, large, heavy branches and
a trunk that tends to be straight and clean.

The bark is pale fawn-gray when young, becoming
fissured with age. The leaves are pinnate and up to 12 in
(30 cm) long. They consist of up to 12 pairs of ovate-
oblong or lanceolate leaflets, which are rich green on the
upper surface, with a slightly paler underside. Distinctive
black leaf buds appear in winter. The fruit sometimes
persist on the tree through winter.

◌ **Golden ash (*Fraxinus excelsior* 'Jaspidea')** Unlike
the common ash, this tree has golden yellow young shoots
and yellowish branches, which are especially obvious in winter.

leaflet up to 4 in (10 cm)
long, 2 in (5 cm) wide

creamy white
flowers

roundly toothed
leaflet margin

PROFILE

↕ Up to 70 ft (20 m)

↔ Up to 50 ft (15 m)

❧ Deciduous

∅ Opposite compound

⊙ S. Europe, S.W. Asia

Fluffy clusters
Borne in large, fluffy, open
clusters (panicles), the
flowers are upright at first,
drooping in late spring.

FEATURES

young
green fruit

Fruit Single, flat, winged
seed, up to 1 ½ in (4 cm) long;
ripening to pale brown

Fraxinus ornus (Oleaceae)

FLOWERING ASH

This broadly spreading, medium-sized tree is
native to southern Europe and southwestern Asia.
It has been widely cultivated in parks, gardens, and
arboreta throughout Europe since before 1700. Unlike
most ash trees, flowering ash produces large panicles of
creamy white, fragrant flowers in spring. Manna sugar,
a form of sweetener tolerated by diabetics, is derived
from the sap of this tree.

The bark is fawn-gray and smooth, even in maturity.
Winter buds are brown and shaped like a bishop's miter.
The leaves are pinnate, up to 8 in (20 cm) long, and consist
of five to nine oblong-ovate leaflets. Matte green on the
upper surface, the leaflets have a pale green underside.
In fall, they turn bright yellow before being shed. Fruit
are borne in long, pendulous clusters and remain on
the tree late into fall.

♤ **Himalayan manna ash (*Fraxinus floribunda*)** This species
has larger flower panicles than the flowering ash and larger leaflets
that are up to 5 in (12 cm) long, with a shiny upper surface.

ovate to lanceolate leaflet, up to 5 in (12 cm) long

PROFILE

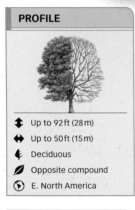

- ↕ Up to 92 ft (28 m)
- ↔ Up to 50 ft (15 m)
- 🍂 Deciduous
- ⬭ Opposite compound
- ◉ E. North America

terminal leaflet

leaflet tapers to a point

Sparsely toothed leaflets
The leaves are pinnate, up to 14 in (35 cm) long. They consist of up to nine short-stalked (except for the terminal leaflet), sparsely toothed leaflets.

FEATURES

Bark Gray, with prominent, narrow fissures, which become deep in maturity

Fraxinus americana (Oleaceae)

WHITE ASH

Also known as American ash, this large, broadly columnar tree is native to eastern North America, from Nova Scotia to Texas and Florida. It was introduced into cultivation in 1724 and is now planted in parks and cities throughout North America and in parts of Europe. The tree is highly prized for its strong, white timber, which is used to make mallets, baseball bats, hockey sticks, and rowing oars.

The winter leaf buds are rusty brown and borne on olive-green shoots. Leaves are dark green on the upper surface and distinctively pale green to almost white beneath. Small, purple male and female flowers appear in clusters on separate trees in spring, before the leaves emerge. The fruit is a 2 in- (5 cm-) long winged seed, ripening from green to brown. It is carried in dense, pendulous clusters throughout fall and early winter.

◊ **Green ash (*Fraxinus pennsylvanica*)** This tree has pinnate leaves with up to 11 leaflets, each up to 6 in (15 cm) long. The leaflets are green, rather than white, on the underside.

*leaflet up to 4 in
10 cm) long, ½–¾ in
(1–2 cm) wide*

Sharp, uneven serrations
The pinnate leaves, up to 10 in
(25 cm) long, consist of up to 13
lanceolate, jagged-toothed, and
widely spaced leaflets.

*glossy dark
green leaflet
upper surface*

PROFILE

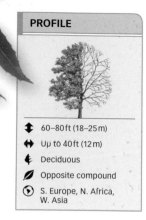

⬍ 60–80 ft (18–25 m)

↔ Up to 40 ft (12 m)

🌢 Deciduous

⌀ Opposite compound

◉ S. Europe, N. Africa,
W. Asia

FEATURES

Bark Gray-brown, displaying
vertical fissures

Fraxinus angustifolia (Oleaceae)

NARROW-LEAVED ASH

This broadly columnar tree has the narrowest
leaves among the members of the *Fraxinus* genus.
It was introduced into cultivation around 1800. There
are several distinctive cultivars related to this species,
including 'Raywood,' which has leaves that turn an
attractive plum-purple in fall.

The shoots are olive-green; winter buds are dark brown
and covered in a dense down. The leaves are similar to
those of the common ash (p.300), but are narrower and
hang downward from the shoots. They are slightly paler
and without hairs on the underside. Small, relatively
inconspicuous, green to purple-green flowers are borne in
clusters in early spring before the leaves unfurl. The fruit
are flattened, winged seeds, each up to 1½ in (4 cm) long.
They are borne in hanging, fawn-colored clusters, which
persist well into winter.

♤ **Pallis's ash (*Fraxinus pallisiae*)** This tree's pinnate leaves
have only seven leaflets. These are more triangular and less shiny
than those of the narrow-leaved ash.

PROFILE

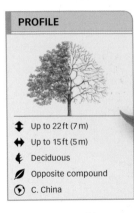

- Up to 22 ft (7 m)
- Up to 15 ft (5 m)
- Deciduous
- Opposite compound
- C. China

Pinnate leaves
The leaf is composed of five, occasionally seven, leaflets borne on purple-green leaf stalks.

dark green upper surface of leaflet

leaflet up to 4 in (10 cm) long, 2 in (5 cm) wide

FEATURES

Bark Light gray; smooth, even in maturity

gray bud

Buds Light gray; covered with fine down in winter

Flowers Creamy white; borne in loose clusters

Fraxinus sieboldiana (Oleaceae)

CHINESE FLOWERING ASH

This slow-growing tree from central China was previously named *Fraxinus mariesii*. It was introduced into western cultivation in 1878, when English plant collector Charles Maries sent it from China to nurserymen Veitch & Sons. Although beautiful when in flower, this broadly spreading tree is uncommon in cultivation and rarely seen in parks and gardens.

It has a round-headed, bushy form in maturity. The leaves are up to 7 in (18 cm) long, and have leaflets that are dark green on the upper surface and silver-green on the underside. They turn yellow and gold before being shed in fall. The flowers emerge in late spring. The fruit are deep purple, single-winged samaras, borne in late summer. They ripen to a light brown color.

○ **Chinese ash (*Fraxinus chinensis*)** A larger, more vigorous tree than the Chinese flowering ash, this tree, up to 50 ft (15 m) tall, bears plump, brownish black winter buds covered in light gray hairs.

PROFILE

- Up to 70 ft (20 m)
- Up to 40 ft (12 m)
- Deciduous
- Opposite simple
- S.E. United States

flowers borne in upright panicle

leaf up to 10 in (25 cm) long, 6 in (15 cm) wide

broadly ovate leaf

Trumpet-shaped flowers
Two-lipped, white flowers with yellow and purple spots are borne in large, open panicles in mid- to late summer.

FEATURES

Bark Light gray-brown; loose, flaking in patches in maturity

seed pod up to 12 in (30 cm) long

Fruit Green seed pods ripening to brown; stay on tree for one year

---

Catalpa bignonioides (Bignoniaceae)

CATALPA

This medium-sized tree is also known as Indian bean tree, after Native Americans, rather than the Indian subcontinent. Native Americans used to dry, paint, and then wear the seeds of this species as decoration. A broadly spreading and highly ornamental tree, it is one of the last to bloom, waiting until mid- to late summer to produce its orchidlike flowers. The catalpa is common in towns and cities throughout the northern hemisphere, mainly due to the fact that it tolerates localized atmospheric pollution very well.

The lax, soft leaves are heart-shaped at the base, and run to a tapered point. They are rarely lobed. Bronze-colored in spring, they quickly fade to grass green on the upper surface, paler on the underside with an occasional covering of hairs. The flowers are white with yellow and purple markings.

◊ **Golden-leaved catalpa (*Catalpa bignonioides* 'Aurea')**
This tree has golden yellow leaves, relatively sparse panicles of flowers, and a smaller stature than the catalpa.

PROFILE

leaf up to 6in (15cm) long, 5in (12cm) wide

flowers borne in dense clusters

↕ Up to 70ft (20m)

↔ Up to 40ft (12m)

🍂 Deciduous

🍃 Opposite simple

◉ W. China

long, slender point at leaf tip

lax petiole

Pink flowers
Delicate, bell-shaped, pink flowers are borne in dense clusters in midsummer.

FEATURES

Bark Dark gray, becomes scaly; peels to reveal gray-pink, fresh bark beneath

Variant Seed pod variant from subspecies 'Duclouxii'

Catalpa fargesii (Bignoniaceae)

FARGES' CATALPA

This broadly columnar, medium-sized, deciduous tree was named after the French Jesuit missionary Père Paul Guillaume Farges, who discovered it in western China in 1896. It is now widely grown in both North America and Europe for its delicate pink flowers, which appear in early to midsummer, several weeks earlier than those on the tree's American cousin, the *Catalpa bignonioides* (p.305).

The leaves are lax and soft, broadly ovate, with a small side lobe that may or may not be present. Borne singly on a long petiole (stalk), the leaves are bronze-colored when they first emerge from the bud. By late spring, they become deep green, with a slight sheen on the upper surface, and paler on the underside. The light pink flowers have maroon spots and yellow blotches at the entrance to the flower throat. Around 15 flowers are borne in large, open panicles, which are up to 6in (15cm) long. These are followed by pendulous seed pods up to 18in (45cm) long. Slender, green, ripening to dark brown, they appear in summer, persisting on the tree till fall.

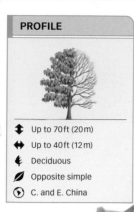
Upright panicles
Open, erect panicles up to
12in (30cm) tall, bear trumpet-
shaped flowers in late spring.

brown, felted
flower bud

blotchy yellow
on inner surface

flower up to 2in
(5cm) long

lilac to pale-purple
on outer surface

FEATURES

Leaves Large, heart-shaped
at base; with a large, shallow
lobe on either side

Fruit Green, pointed,
egg-shaped capsule ripening
to woody brown

Seeds Winged, borne in green
fruit, dispersed by wind

Paulownia tomentosa (Scrophulariaceae)

EMPRESS TREE

Also known as the princess tree, or the foxglove tree,
this fast-growing, medium-sized, deciduous tree was
introduced into western cultivation in 1834. Since then,
the empress tree has become popular for planting in
parks, gardens, and arboreta. It takes its genus name from
Anna Pavlovna, the daughter of
Czar Paul I of Russia.

The bark is gray and smooth,
even in maturity, similar to that
of the common beech (p.152).
The leaves are ovate, up to 18in
(45cm) long and wide—even
larger on coppiced regrowth.
They are lax and soft, dark green
on the upper surface, lighter on
the underside, and covered
in fine, sticky hairs. Fragrant,
tubular flowers are borne in
upright open panicles.

Columnar habit
The empress tree is rounded
to broadly columnar with
very stout shoots that are
hairy when young.

pea-sized fruit

Ovate leaflets
The large, pinnate leaves of this tree consist of broadly ovate leaflets running to a slender point.

leaflet tapers to a point

FEATURES

Bark Light gray-brown, smooth; becomes vertically fissured in maturity

Rhus verniciflua (Anacardiaceae)

VARNISH TREE

Also known as Chinese lacquer tree, this fast-growing, medium-sized, deciduous tree has been cultivated in Europe since before 1862. Broadly conical when young, it develops a rounded habit in maturity. Its name refers to the fact that its sap is the source of the varnish used to give a high-gloss finish to Chinese and Japanese lacquerware. In China, oil is also extracted from the fruit to make candles. However, all parts of the tree are poisonous and contact with skin can cause irritation and blistering.

The varnish tree's branches are long and gangly, and sparsely covered with large, pinnate leaves up to 24 in (60 cm) long. These consist of 9 to 19 leaflets, each up to 6 in (15 cm) long. They are rich, glossy green on the upper surface, and paler with a fine covering of hairs on the underside. In summer, small, yellowish white flowers are produced in large, pendulous panicles. These are followed by clusters of berrylike, yellowish green fruit.

Flowers in long panicles
Chinese sumach bears small, creamy yellow flowers, carried in conical panicles, in summer.

flower panicle up to 8 in (20 cm) long

lanceolate to oblong leaflet, up to 5 in (12 cm) long

PROFILE

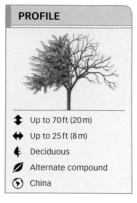

- ↕ Up to 70 ft (20 m)
- ↔ Up to 25 ft (8 m)
- Deciduous
- Alternate compound
- China

Rhus potaninii (Anacardiaceae)

CHINESE SUMAC

Also known as Potanin's sumac, this tree is named after the Russian explorer and botanist Grigori Potanin, who discovered and recorded the species in northern China. The broadly spreading, medium-sized tree has been cultivated in the west for at least 100 years and is found in parks, gardens, and arboreta throughout Europe. In the wild in China, it has a neat, round-topped shape. However, in cultivation, it is likely to be multi-stemmed and spreading.

The bark is gray-pink and smooth. The long and spreading branches are bare except for the ends, which carry large, pinnate leaves. The leaves, up to 14 in (35 cm) long, are borne on hairy shoots and carry 7 to 11 leaflets, which run to a fine point. The leaflets are dark green on the upper surface, paler on the underside, and usually turn bright red or pink-purple in fall. The tree bears flowers in summer, and these are followed by rounded, green, berrylike fruit, which are covered in fine hairs. The fruit turn red when ripe.

FILE

Leaves and fruit
Breadfruit leaves are large, with six to nine pointed lobes. The green fruit, borne singly or in clusters of two or three at ends of branches, is rounded to football shaped.

leaf up to 36 in (90 cm) long

- Up to 80 ft (25 m)
- Up to 70 ft (20 m)
- Evergreen
- Alternate simple
- Malaysia, Indonesia, Pacific Ocean islands

FEATURES

Bark Gray, smooth; releases sticky white latex

Fruit Green, rounded to oval; up to 8 in (20 cm) long with warty skin

Artocarpus altilis (Moraceae)

BREADFRUIT

This fast-growing tree is native to Malaysia, Indonesia, and the Pacific Ocean islands. However, it has been cultivated in other parts of the tropics for centuries. Broadly columnar, it has ascending, arching branches and a dense crown of foliage. When cooked, its fruit tastes like freshly baked bread. The most famous transportation of this plant was by Captain Bligh aboard HMS *Bounty* in 1789. He was tasked with transporting breadfruit plants from Tahiti in the Pacific to the West Indies. The lavish attention paid to the plants rather than the sailors caused a mutiny along the way.

The rough, ovate leaves are deep glossy green on the upper surface, paler on the underside. Tiny male and female flowers are borne in profusion on the same tree. They appear on a capitulum, which is a flower head like a false fruit. The actual fruit weighs up to 11 lb (5 kg). Its green outer skin has hexagon-shaped discs. Inside, the milky flesh looks a little like that of a melon. It is starchy when unripe and usually fragrant when ripe.

Jackfruit foliage
The leaves are dark green with a slight sheen on the upper surface and pale green with some down on the underside.

leaf up to
8 in (20 cm) long

oblong to
egg-shaped leaf

PROFILE

- Up to 70 ft (20 m)
- Up to 70 ft (20 m)
- Evergreen
- Alternate simple
- S.E. Asia; known only in cultivation

FEATURES

Bark Relatively smooth, reddish brown to gray

Fruit Very large, up to 36 in (90 cm) long, 20 in (50 cm) wide

Artocarpus heterophyllus (Moraceae)

JACKFRUIT

The jackfruit is believed to have originated from southern India, but has been so widely cultivated throughout Asia that it is now difficult to pinpoint its natural range. Rather like breadfruit (p.310), the fruit of the jackfruit tree are cultivated for their edible flesh. The tree is fast-growing, broadly columnar, and medium-sized, with a single trunk that branches into a dense crown of foliage.

The mature, dark green leaves are leathery. Both male and female flowers are borne on the same tree. Small and green, the flowers emerge directly from the trunk and main branches. They are followed by rounded to oval fruit, weighing up to 44 lb (20 kg), which are borne on woody stalks directly attached to the trunk and main branches. The fruit has waxy flesh inside, which may be pink or golden yellow with a strong, unpleasant odor, but a sweet taste rather like sour banana. The jackfruit flesh is a nutritious carbohydrate used in many dishes, from custards to curries.

↕ 50–100ft (15–30m)

↔ Up to 100ft (30m)

🌿 Semi-evergreen

🍃 Alternate simple

⟳ Southeast Asia

Cordate leaves
Bo leaves are heart-shaped,
with an elongated, slightly
curved tip.

*leaf up to 8in (20cm)
long, 5in (12cm) wide*

*pale yellow
leaf midrib*

*flexible leaf stalk, up
to 4in (10cm) long*

FEATURES

Bark Smooth, gray; develops
vertical fluting and buttressing
in maturity

Ficus religiosa (Moraceae)

BO TREE

According to Buddhist beliefs, the Buddha attained
enlightenment under the bo tree, also known as the
sacred fig. This long-lived species (some specimens
are believed to be 2,000 years old) is sacred to both
Buddhists and Hindus, and is usually grown close to
temples. Its native habitat extends from India and
Pakistan to Indonesia, but it is widely planted outside
these countries, including subtropical areas of the US.

A wide-spreading tree, its trunk may extend up
to 10ft (3m) in diameter in maturity. With age, some
surface roots start radiating from the trunk's wide
base. Aerial roots may also hang down from the
branches. The leaves range in color from bright
green to blue-green, with a pale yellow midrib and
veining. The flowers are red but very small. Both
male and female flowers are borne on the same tree.
The fruit is a small fig, up to 1/2in (1.5cm) long, green
ripening to purple. It is borne in rounded and stalkless
clusters along the twigs.

glossy leaf, up to 10 in (25 cm) long

pale yellow leaf midrib

PROFILE

↕ 70–80 ft (20–25 m)

↔ Up to 70 ft (20 m)

🌿 Evergreen

◪ Alternate simple

◉ India

Distinctive leaves
The leaves are leathery, elliptical, and blunt-tipped. They have prominent pale yellow veining.

FEATURES

Aerial roots Grow down to the ground; thicken once they reach the soil

Ficus benghalensis (Moraceae)

BANYAN TREE

The banyan is one of the oldest and largest trees on Earth, with specimens more than 5,000 years old still in existence. One banyan in the botanic garden near Kolkata (Calcutta), India, has a crown circumference approaching 0.6 miles (1 km). As with other fig trees, banyans are stranglers, starting life as a seed deposited by a bird among the branches of another tree. This seed develops aerial roots that eventually strangle the host tree. The tree's low-spreading branches grow more aerial roots, some of which develop into secondary trunks—old trees can have hundreds of such trunks. The canopy of the banyan tree casts a dense shade.

The bark is pale gray and relatively smooth. The leaf buds are covered by two scales, which fall as the leaves grow. The leaves are bronze-pink when young, turning dark green in maturity. The banyan produces numerous small flowers. The fruit is a spherical fig, up to ¾ in (2 cm) wide, borne in pairs in the leaf axils. It is green and ripens to scarlet.

leaf up to 4 in (10 cm)
long, 2 in (5 cm) wide

Smooth, leathery leaves
Ovate to elliptic leaves are glossy
green on the upper surface, paler
on the underside, and thinner in
texture than most fig leaves.

PROFILE

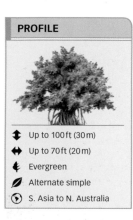

- ↕ Up to 100 ft (30 m)
- ↔ Up to 70 ft (20 m)
- ♣ Evergreen
- ⬭ Alternate simple
- ◉ S. Asia to N. Australia

Ficus benjamina (Moraceae)

WEEPING FIG

In cool regions of the world, this tree is
commonly grown as a houseplant. However, in
the tropics it is a broadly spreading, eventually
weeping, large evergreen tree. It is native to south and
southeast Asia, north Australia, and the southwest
Pacific but has been widely cultivated throughout
the tropics. The weeping fig is a graceful tree with
strongly ascending branches that weep at the tips.
Like most figs, it is a strangler, developing aerial
roots that descend to form a dense latticework
around the main trunk.

The bark is pale gray, sometimes almost white,
and exudes white latex. The alternate leaves
shimmer in the slightest breeze. Green-cream
flowers are small and not very visible. The fruit are
small rounded figs, which are borne in pairs in the
leaf axils nearest the ends of the twigs. They ripen
from green to pink, then turn scarlet before ripening
to purple, and eventually black.

leaf 6–12 in
(15–30 cm) long

Large, elliptic leaves
This fig tree bears alternate leaves,
dark green on the upper surface,
golden brown on the underside,
which bleed a milky sap when cut.

PROFILE

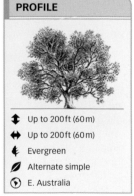

↕ Up to 200 ft (60 m)
↔ Up to 200 ft (60 m)
🌿 Evergreen
🍃 Alternate simple
◉ E. Australia

Ficus macrophylla (Moraceae)

MORETON BAY FIG

This broadly spreading tree is indigenous to eastern
Australia, from Queensland in the north to New South
Wales and Lord Howe Island in the south. It is widely
cultivated as an ornamental tree across warmer parts of
the world, including the Mediterranean and California.
It is known as a strangler because its seedling germinates
in the canopy of a host tree and lives as an epiphyte—
growing on another tree non-parasitically— until its aerial
roots touch the ground.

The bark is light brown to
gray, smooth, developing
heavy buttressing and surface,
snakelike roots in maturity. The
small flowers require pollination
by a specific wasp, known
simply as the fig wasp. The fruit
is a fig, up to 1 in (2.5 cm) wide,
green ripening to purple with
lighter spots

Buttressed trunk
The Moreton Bay fig's
massive base and equally
large spread make it suitable
for wide open spaces.

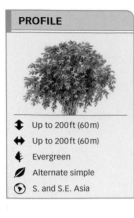
prominent midrib

leaf up to 12 in (30 cm) long, 6 in (15 cm) wide

Glossy green leaves
The thick leaves of the Indian rubber tree mature from pink to dark green, turning glossy and smooth, with a leathery texture.

Ficus elastica (Moraceae)

INDIAN RUBBER TREE

Native to the eastern Himalayas, northeast India, Burma, and Malaysia, this large, broadly spreading fig is as wide as it is tall. It was the original source of rubber until the Brazilian *Hevea brasiliensis* rubber tree was discovered. As with most tropical figs, this tree is a strangler. It develops secondary trunks from aerial roots, and has long, sinuous, surface roots, which splay out for several yards from the main trunk.

The oblong to elliptic leaves have a single, distinct midrib and a pointed tip. They are arranged spirally on the shoots, which are at the ends of long, bare branches. The fruit is an oval, stalkless fig, up to ¾ in (2 cm) long, borne in pairs in the leaf axils toward the ends of the twigs. The greenish yellow figs are produced all year round. The tree begins to bear fruit from around 20 years of age.

Signature foliage
The leaves are tough, waxy, bright green with a slight sheen on the upper surface. The fruit are green, rounded, and pendulous.

fruit ¹⁄₈–¹⁄₄ in (4–5mm) wide

fruit open at one end

thick, obovate leaf, up to 8in (20cm) long

PROFILE

- ↕ Up to 50ft (15m)
- ↔ Up to 50ft (15m)
- 🗲 Evergreen
- ⬯ Opposite simple
- ⊙ Caribbean, Florida Keys, S.E. Mexico

Clusia rosea (Clusiaceae)

AUTOGRAPH TREE

This tree thrives in coastal regions, being tolerant of salt spray, high winds, and saline soil. Its common name refers to its leaves, which are so thick that words can be carved on to them—they have even been used to make playing cards. A strangler, the medium-sized, spreading autograph tree begins its life as an epiphyte. Once attached to the ground, it quickly develops a sprawling habit, with horizontal branches and a dense crown.

The bark is light gray and smooth. The leaves have a bright green upper surface that contrasts with their pale green underside. Flowers are 4in (10cm) wide, consisting of seven fleshy, white petals surrounding a cluster of green-yellow stamens. They are short-lived, opening in the late afternoon and beginning to turn brown and die by the following morning. The flower buds are globular, mostly white with a pink blush. The fruit are glossy green, globular capsules up to 2in (5cm) wide. When mature, they split to release small, orange-white seeds.

deep green upper surface of leaf

leaf up to 10 in (25 cm) long

Cacao leaves
The leaves are oblong to elliptic, with a short pointed tip. Leathery and smooth, they have a pale underside.

PROFILE

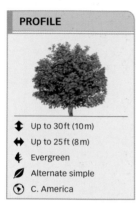

↕ Up to 30 ft (10 m)

↔ Up to 25 ft (8 m)

🌿 Evergreen

✦ Alternate simple

◐ C. America

FEATURES

Bark Rich brown, relatively smooth throughout life

pod up to 8 in (20 cm) long

Fruit Ovoid; ripening from green to orange and maroon, then brown

Theobroma cacao (Malvaceae)

CACAO

This small, spreading tree provides cocoa—the key ingredient of chocolate. Cacao is indigenous to Central America, but has been widely cultivated in the West Indies, Africa, Java, and Sri Lanka for its seeds—cocoa beans—which are roasted and ground to produce cocoa butter.

It has alternately arranged, large pendulous leaves. The creamy yellow to pink flowers, up to ½ in (1 cm) wide, each have five petals. They emerge directly from the trunk and larger branches in spring, and sporadically at other times of the year too. These are followed by the fruit—a pod, shaped like a football and marked by longitudinal ribbing. These are attached to the trunk and main branches by short, woody stalks. They contain up to 100 purple cocoa beans, up to ¾ in (2 cm) long, each set in a pleasantly flavored, yogurt-like pulp.

♤ **Cupuacu (*Theobroma grandiflorum*)** This tree is easily distinguished by its larger, oval fruit, which are covered in a brown, "fuzzy" suede coat, and larger leaves, up to 14 in (35 cm) long.

PROFILE

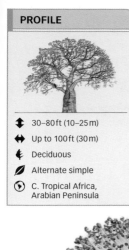

- 30–80 ft (10–25 m)
- Up to 100 ft (30 m)
- Deciduous
- Alternate simple
- C. Tropical Africa, Arabian Peninsula

white flower, up to 6 in (15 cm) wide

yellow-tipped stamen

Showy white flower
Baobab flowers have crinkled petals and numerous yellow or purple stamens. The flowers hang singly from branches on long cords.

FEATURES

Bark Mid-gray, smooth

Fruit Brown, pitcher-shaped; up to 12 in (30 cm) long, 4 in (10 cm) wide

Adansonia digitata (Malvaceae)

AFRICAN BAOBAB

This is the most widespread of the eight species of *Adansonia* (baobabs), and the only species native to central tropical Africa and parts of the Arabian Peninsula. The majority of the other species are native to Madagascar. Several of African baobab's largest specimens are believed to be more than 1,000 years old.

The African baobab is a large tree, with a massive trunk that may attain a circumference of 100 ft (30 m). The leaves are glossy dark green and rounded. They are divided into five to nine leaflets, each up to 6 in (15 cm) long, with a pale green midrib. The tree may remain bare for more than half the year. Its flowers open at night to release a pungent scent like that of overripe melon, which is attractive to fruit bats—the tree's primary pollinator. The fruit hang on long stalks; they contain up to 30 seeds, surrounded in an acidic, edible pulp.

Madagascan baobab (*Andansonia madagascarionsis*)
Unlike the African baobab, this tree is native to Madagascar. It is also smaller in stature and bears dark red flowers.

PROFILE

↕ Up to 200ft (60m)

↔ Up to 70ft (20m)

🍂 Semi-evergreen

🌿 Alternate compound

🌐 Central and South America; Africa; S. Asia

leaflet broadest below its centre

leaflet up to 6in (15cm) long

leaflet tapers to a narrow tip

Long-stalked leaves
Kapok leaves are up to 12in (30cm) wide, and divided into five to nine lanceolate leaflets. The leaves are held on long, slender stalks.

FEATURES

Bark Silver-gray, often covered with thick spines

fluffy cream fiber

Fruit Banana-shaped brown pod, up to 6in (15cm) long

Ceiba pentandra (Malvaceae)

KAPOK

This tree occurs naturally in Central and South America, Africa, and southern Asia. This shows that it existed 250 million years ago when all three continents were interconnected and made up part of the supercontinent known as Pangea. By far the highest population of kapok is in Central America, where the ancient Mayan civilization considered the tree to be sacred. It is a large, broadly conical, semi-evergreen tree, reaching up to 200ft (60m) in height.

The kapok's trunk base is heavily buttressed in mature specimens. In spring, this tree bears small, fragrant flowers in clusters. These flowers, up to ¾in (2cm) wide, have five pale yellow, waxy petals, which curve back on themselves to reveal five extended stamens. The flowers are followed by long, leathery fruit pods, which split open vertically to release seeds surrounded by masses of fiber. The fiber, also known as kapok, was once used as a filling for, among other things, pillows and life jackets.

PROFILE

- Up to 70 ft (20 m)
- Up to 70 ft (20 m)
- Deciduous
- Alternate compound
- South America

Silk floss flowers
The flowers are bright rose-pink with five petals. The inner side of each petal usually becomes white toward the base.

purple markings on inside of petal

long, curved, pink pistil, initially enclosed by petals

FEATURES

Bark Gray-green; pockmarked with thick, blunt thorns

Chorisia speciosa (Malvaceae)

SILK FLOSS TREE

Native to South America, the silk floss tree, also called *Cebia speciosa*, is cultivated elsewhere as an ornamental tree in parks and gardens. This fast-growing, broadly spreading, medium-sized tree is attractive while in flower.

The bark is marked with blunt thorns. The density of the thorns varies from tree to tree, and thorns may appear on larger branches as well. The base of the trunk may become distended in maturity. The green, compound leaves are up to 10 in (25 cm) wide, and consist of five to seven narrowly elliptic leaflets, each up to 5 in (12 cm) long. The leaf margins are irregularly toothed, and leaflets are borne on a long leaf stalk. Hibiscus-like flowers emerge from bulging green buds in the leaf axils from spring into summer, and grow up to 6 in (15 cm) wide. These are followed by ovoid to pear-shaped woody pods, 8 in (20 cm) long. The pods contain black seeds surrounded in fluffy material, which resembles sheep's wool—rather like the kapok (p.320).

PROFILE

Up to 130 ft (40 m)

Up to 100 ft (30 m)

Semi-evergreen

Alternate simple

Tropical South America

Flower cluster
Pale yellow flowers are borne above the leaves, in panicles up to 4 in (10 cm) long. Each flower has six unequal-sized petals.

flower 1 in
(2.5 cm) wide

bud opens to
six-petaled flower

FEATURES

Bark Light gray; smooth when young, fissures vertically in maturity

Leaves Narrowly oblong, dark green, leathery

Fruit Brown, hard-shelled, woody, 4–5 in (10–12 cm) wide

Bertholletia excelsa (Lecythidaceae)

BRAZIL NUT

As the name suggests, this large, semi-evergreen tree is native to Brazil, but it also inhabits other countries bordering the Amazon basin. It is a typical emergent rainforest tree—standing above the canopy with a long, straight trunk devoid of branches for two thirds of its length and topped by a crown made up of long branches with dark foliage. The tree's trunk exceeds 6 ft (2 m) in width.

The bark, on maturity, resembles that of a mature common ash (p.300). The leaves are up to 14 in (35 cm) long and 6 in (15 cm) wide, each with pronounced veining and a pale green central midrib. In the dry season, they turn bronze-red before falling. The fruit is shaped like a cannonball with a flattened top. Each fruit may weigh up to 4 ½ lb (2 kg). Inside are the hard-shelled seeds known as Brazil nuts.

PROFILE

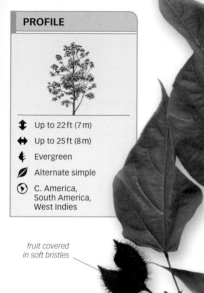

⬍ Up to 22 ft (7 m)

⬌ Up to 25 ft (8 m)

🍂 Evergreen

🌿 Alternate simple

📍 C. America,
South America,
West Indies

*leaf up to
8 in (20 cm) long*

*prominent
leaf vein*

*fruit covered
in soft bristles*

Fruit and leaves
The annatto fruit is an egg-shaped
husk. Leaves are bright green and
broadly ovate.

FEATURES

Flowers Cup-shaped, pale
pink; borne in clusters

Seeds Small; exude bright red
color when crushed

Bixa orellana (Bixaceae)

ANNATTO

This small, broadly spreading tree has a multi-stemmed, shrublike habit, and is native to Central and South America as well as the West Indies. It is a popular plant for cultivation in tropical gardens because of its attractive flowers and useful seeds. The annatto is best known as the source of a bright red dye, which comes from the seeds. The dye obtained is used to color cosmetics, such as lipstick, and food, such as smoked fish and red-veined cheese.

The bark is light gray-brown and smooth. Leaves are faintly glossy, growing up to 6 in (15 cm) wide. The flowers are clustered at the ends of twigs on smaller branches. They are up to 2 in (5 cm) wide, with five petals surrounding a central cluster of pink stamens. The fruit is a husk that ripens from yellow to red, borne in clusters of four to five at the ends of twigs and can grow up to 3 in (7 cm) long. Each husk contains up to 50 red seeds.

PROFILE

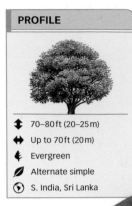

- 70–80 ft (20–25 m)
- Up to 70 ft (20 m)
- Evergreen
- Alternate simple
- S. India, Sri Lanka

bright green leaf upper surface

glossy leaf, up to 6 in (15 cm) long, 2 in (5 cm) wide

wavy-edged leaf margin

Vibrant green foliage
Ebony leaves are borne alternately on stems. The leaves are paler with dark markings on the underside.

FEATURES

Bark Dark gray; peels off in rectangular flakes in maturity

Diospyros ebenum (Ebenaceae)

EBONY

There are several trees with the commercial name ebony, but this species—with its dense black, extremely hard heartwood—is the best known. It is native to southern India and Sri Lanka. Ebony is most valued for its timber, which has a wide variety of uses, from woodturning and carving to furniture making. In the wild, it is a slow-growing, medium-sized, evergreen tree, reaching about 70–80 ft (20–25 m) tall.

The tree has oblong-elliptic leaves, with blunt tips. Although male and female flowers are borne on separate trees, they are similar in appearance. They are small, up to ¾ in (2 cm) wide, with four dull white, thick petals. The female flowers are borne singly, while the male flowers bloom in clusters of up to 15. The female flowers are followed by the fruit, which is an edible rounded berry—similar in appearance to a small persimmon. It is ¾ in (2 cm) wide, green when young, ripening to black, and held in a green woody cup that was the old flower sepal.

dark glossy green leaf

7–13 veins on leaf

leaf up to 28 in (70 cm) long

bright green stem, up to 24 in (60 cm) long

Papaya leaves
Large leaves are borne on lax leaf stems. They are palmately divided into deeply cut lobes.

PROFILE

- Up to 22 ft (7 m)
- Up to 12 ft (4 m)
- Evergreen
- Alternate simple
- Central and South America

FEATURES

flower up to ¾ in (2 cm) wide

Flowers Male and female borne on separate trees

Fruit Fleshy; green ripening to orange-yellow

Carica papaya (Caricaceae)

PAPAYA

This fast-growing, short-lived, small, columnar evergreen is native to Central and South America but is also grown elsewhere for its juicy, edible fruit. Although it takes on treelike proportions, the papaya is not strictly a tree but a long-lived, herbaceous plant. It displays a single stem with spirally arranged leaves that grow only from the top. The trunk does not lay down wood, as in a normal tree, and always remains soft.

Papaya bark is light brown and regularly patterned with scars where old leaf stems were attached. Toward the crown and growing tip, the stem color is bright green. Dark glossy green leaves have five to seven lobes, each with further incisions. The small, waxy flowers are similar to frangipani (*Plumeria*). Cream to greenish white in color, they appear in the leaf axils, and develop into a pear-shaped fruit. Around 6–18 in (15–45 cm) long, the fruit hangs close to the stem and consists of yellowish orange pulp and black seeds.

deep cleft divides
leaf apex into two
rounded lobes

deep olive-green leaf,
up to 5 in (12 cm) long,
6 in (15 cm) wide

Orchidlike flower
Camel's hoof tree has lilac-pink,
five-petaled flowers, resembling
an orchid. The flowers bloom in
clusters on short stalks.

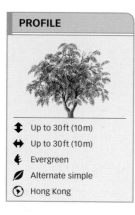

PROFILE

- ↕ Up to 30 ft (10 m)
- ↔ Up to 30 ft (10 m)
- ❀ Evergreen
- ⬭ Alternate simple
- ⊙ Hong Kong

Bauhinia x *blakeana* (Fabaceae)

CAMEL'S HOOF TREE

Shaped like a camel's hoof, the distinctive leaves of this tree aid in its identification. This tree is also known as the Hong Kong orchid tree, and is widely cultivated across the tropics and subtropics. It is a hybrid between the orchid tree (*B. variegata*) and the purple camel's hoof tree (*B. purpurea*), and was discovered near the ruins of a Hong Kong house around 1880. It is believed that all the trees of this species in cultivation today are derived from one plant, which was propagated from the 1880 specimen and then grown in the Hong Kong botanic garden.

This fast-growing, medium-sized tree has a broadly spreading habit, often with multiple stems. It has a smooth, gray-green bark. The leaves are orbicular (disc-shaped), thick, and leathery. The tree bears highly fragrant flowers, up to 6 in (15 cm) wide, borne in clusters of two to three at the tips of branchlets. Flowers have deep burgundy or purple streaks on just one of their five petals.

Showy flowers
When the royal poinciana is in bloom, the profusion of its bright vermilion flowers almost obscures the delicate, bipinnate foliage.

widely spread, vermilion petal

⬍ Up to 70 ft (20 m)

⬌ Up to 70 ft (20 m)

🌡 Semi-evergreen

⬤ Alternate compound

⦿ Madagascar

FEATURES

Fruit Hang down en masse from branches; green at first, ripening to brown

Delonix regia (Fabaceae)

ROYAL POINCIANA

This broadly-spreading tree originates from Madagascar, where it is now endangered in the wild. However, the royal poinciana is widely cultivated across tropical and subtropical regions of the world as an ornamental tree. It is also known as the flame tree or flamboyant tree, and in India, gulmohar, which translates from Hindi as "peacock flowers," a reference to its showy, bright orange-red flowers.

The royal poinciana has smooth, light brown bark and a short trunk, which quickly breaks into long, horizontal branches. The leaves, up to 24 in (60 cm) long, are frond like, similar to that of the jacaranda (p.333). They have a bright green upper surface and a paler underside. Flowers are borne in spring or early summer. They have five petals, four of which are an intense vermilion, while the fifth, normally the central petal, is streaked with cream, red, and yellow. The flowers are followed by long, flattened, and pendulous seed pods, which grow up to 24 in (60 cm) long and 2 in (5 cm) wide.

↕ 30–80 ft (10–25 m)

↔ Up to 50 ft (15 m)

🌿 Evergreen

🍃 Opposite simple

⊙ Coastal regions of
E. Africa, India, S.E.
Asia, W. Pacific islands,
N. Australia

*seedling up to
30 in (75 cm) long*

Pendulous fruit
The fruit is a dark brown,
leathery, pear-shaped pod.
Seeds germinate within the
pods while still on the tree.

FEATURES

Leaves Shiny, bright green
upper surface, paler underside,
with warty spots

Aerial roots Prop roots arch
down from lower trunk to
anchor tree to seabed

Rhizophora mucronata (Rhizophoraceae)

ASIATIC MANGROVE

This broadly spreading, multi-stemmed evergreen
tree is native to coastal regions of East Africa, India,
southeast Asia, western Pacific islands, and northern
Australia. It grows with its roots in salt water and has
its own built-in desalination mechanism and defenses
against waterlogging.

The tree's bark is light gray. Its aerial roots anchor it
to the seabed, giving the tree a stilted appearance. Its
obovate leaves are up to 7 in (19 cm) long and 4 in (10 cm)
wide. They have a sharply pointed tip and turn from green
to red-brown. Creamy white, four-petaled flowers, up to
¾ in (2 cm) wide, are borne in clusters of three to five in
the leaf axils. The fruit are pods, up to 2 in (5 cm) long, also
borne in the leaf axils. They contain seeds that germinate
into seedlings on the parent tree, growing straight and long
before dropping from the tree.

♤ **Red mangrove (*Rhizophora mangle*)** This tree is recognized
by its more pronounced arching roots, smaller leaves up to 5 in
(12.5 cm) long and 2 in (5 cm) wide, and pale yellow flowers.

fruit up to 1¼in (3cm) long

elliptic leaflet, up to 8in (20cm) long

Leaves and fruit
The trifoliate, leathery leaves are spirally arranged on the shoot. The green, capsulelike fruit ripens to woody brown.

PROFILE

↕ Up to 130ft (40m)

↔ Up to 70ft (20m)

🍂 Semi-evergreen

🌿 Alternate compound

⊙ Brazil, Venezuela

FEATURES

Bark Pale gray-brown; smooth throughout life

Hevea brasiliensis (Euphorbiaceae)

RUBBER TREE

This fast-growing, large, semi-evergreen, broadly columnar tree is indigenous to both the Amazon river basin in Brazil and the Orinoco river basin in Venezuela. It has been widely planted elsewhere in tropical regions, such as in Malaysia, because it is the main source of natural rubber for commercial use.

Rubber can be tapped from the trees when they are six years old. This is done by cutting a slanting groove in the outer bark and allowing the milky white latex to ooze out into a collecting cup. Trees that are no longer tapped retain V-shaped scars on their otherwise smooth bark for many years and are a good identification feature.

The leaves are dark green on the upper surface, paler on the underside. They may turn orange-red and fall in the dry season. Small, greenish white, fragrant flowers appear in the leaf axils in short, erect panicles. The fruit is a smooth, three-lobed capsule. When fully ripe, it bursts open with a loud crack to release the seeds inside.

glossy, deep green
upper surface of leaf

Narrow, lax leaves
Lanceolate leaves of the
mango tree are pale green
on the underside and hang
down on long leaf stalks.

pale yellow
midrib

pronounced veining
on leaf surface

PROFILE

↕ 100 ft (30 m)

↔ 50 ft (15 m)

🌿 Evergreen

🍃 Alternate simple

🜨 India, E. Asia

FEATURES

Bark Smooth gray-brown;
turns dark, furrowed with age

unripe
fruit

Fruit Green ripening to
yellow, orange, or red

Mangifera indica (Anacardiaceae)

MANGO

Native to India, mango has been widely cultivated throughout the tropics for millennia. It was grown in East Asia as early as 500 BCE. Today, it is probably the best known and most popular of all tropical fruit and is regularly eaten worldwide.

The trunk of this tree may become buttressed in maturity. New leaves emerge copper-red and are very lax. The leaves are up to 12 in (30 cm) long and 3 in (7.5 cm) wide, and hang from 4 in- (10 cm-) long leaf petioles (stalks). Flowers are tiny, pale yellow, cream, or pale pink, and clustered in hundreds along terminal flower stalks. Flowers are followed by fruit borne in hanging, stalked bunches. The size of the fruit is variable, but normally similar to that of the avocado (p.111).

Large, dense habit
The largest cultivated fruit tree, the mango has a trunk that carries a dense crown of dark, luxuriant foliage.

PROFILE

- Up to 70 ft (20 m)
- Up to 70 ft (20 m)
- Semi-evergreen
- Alternate compound
- C. America, West Indies

dark green leaflet

leaflet up to 3 in (7 cm) long, 1 ¼ in (3 cm) wide

Sickle-shaped leaflets
The mahogany leaves consist of four to eight pairs of ovate to sickle-shaped leaflets. There is no terminal leaflet.

FEATURES

Bark Dark red-brown, scaly

pod up to 4 in (10 cm) long, 2 ½ in (6 cm) wide

Fruit Oval, woody capsules; contain up to 50 winged seeds

Swietenia mahagoni (Meliaceae)

MAHOGANY

This slow-growing tree is probably best known as the source of one of the world's most valued timbers. Mahogany is a medium-sized, broadly columnar, semi-evergreen tree with upward-sweeping branches. It has dense, durable wood, which has a rich, red color and can be polished to a high shine. It has been favored by furniture makers ever since it was discovered in Central America and the West Indies in the 17th century. Unfortunately, due to overcutting, it is now endangered in the wild, and its timber should only be obtained from trees grown in managed plantations.

The mahogany's trunk may become buttressed with age. The pinnate leaves are up to 10 in (25 cm) long with four to eight pairs of leaflets. Flowers are small, greenish yellow to white, and fragrant, emerging in early summer and persisting on the tree until early fall. They are borne in branched clusters in the leaf axils. The fruit is a woody capsule, stalked and upright with winged seeds.

conspicuous
veins on leaf

ovate-elliptic leaf, up
to 20 in (50 cm) long,
10 in (25 cm) wide

Teak leaves
The tree bears rough, leathery,
and rich green leaves. Each leaf
has a wavy margin.

PROFILE

- ↕ Up to 120 ft (35 m)
- ↔ Up to 70 ft (20 m)
- ♣ Deciduous
- ⬤ Opposite simple
- ◉ India, Indonesia,
  Malaysia, Burma

FEATURES

Bark Pale gray; flakes in
maturity to reveal yellow
patches beneath

Tectona grandis (Lamiaceae)

TEAK

Teak is not only the common name of this broadly
columnar, fast-growing tree, but also of its timber and
wood products. The tree is grown across the world for
its timber and has been naturalized in many tropical
countries, including parts of
Africa and the Caribbean.

The trunk develops vertical
fluting with age. New leaves
emerge in the wet season and
are followed by small, fragrant,
white flowers. The leaves are
borne on thick petioles, up to
1 1/2 in (4 cm) long, and are laden
with soft, white hairs on the
underside. Flowers are carried in
panicles up to 16 in (40 cm) long
and 12 in (30 cm) wide. The fruit
are round, fleshy, and purplish
red, ripening to brown.

Teak timber
Since its introduction into
Europe in the early 1800s,
teak has been widely used
to make furniture.

Trumpet-shaped flowers
Vibrant lilac-blue, trumpet-shaped flowers are carried on erect flower clusters.

flower up to 2 in (5 cm) long

panicle up to 12 in (30 cm) long

bipinnate leaf, up to 18 in (45 cm) long

PROFILE

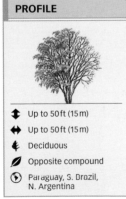

↕ Up to 50 ft (15 m)

↔ Up to 50 ft (15 m)

🍂 Deciduous

▱ Opposite compound

◉ Paraguay, S. Brazil, N. Argentina

Jacaranda mimosifolia (Bignoniaceae)

JACARANDA

One of the best known subtropical trees, the jacaranda is native to South America but has been cultivated in almost every part of the world where there is no risk of frost. It is broadly spreading and medium-sized with an open, branched crown.

The bark is dark gray and smooth when young, becoming scaly in maturity. The leaves are fernlike, with a maximum of 20 pairs of pinnae, each carrying up to 28 pairs of bright green, ovate to elliptic ½ in- (1 cm-) long leaflets. The overall effect is reminiscent of mimosa—hence the species name. Flowers appear in spring and summer. The fruit is a 2 in- (5 cm-) wide, flat brown pod, which contains many winged seeds.

Jacaranda in bloom
The mature tree produces a striking mass of lilac-blue flowers, which last for around two months.

PROFILE

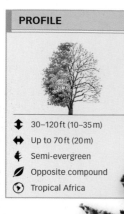

↕ 30–120 ft (10–35 m)

↔ Up to 70 ft (20 m)

⚡ Semi-evergreen

∅ Opposite compound

⊙ Tropical Africa

Tulip-like flowers
This tree bears fragrant blooms that are bright reddish orange on the outside and golden yellow inside. They have frilly, gold-rimmed borders.

curved, pitcher-shaped flower, up to 6 in (15 cm) long

FEATURES

Bark Pale gray, smooth

leaflet up to 4 in (10 cm) long

Leaves Pinnate, up to 24 in (60 cm) long; with 9 to 21 leaflets

Spathodea campanulata (Bignoniaceae)

AFRICAN TULIP TREE

Also known as flame of the forest, this tree is widely grown throughout the tropical and subtropical world in parks and gardens. It is especially popular in subtropical regions of the US, and in India, where it is planted as a street tree. Although considered an evergreen, it sheds its leaves in dry summers.

The African tulip tree carries a few low branches, but develops a thick, dense crown of foliage. The leaves consist of ovate, dark green leaflets, each with a slight sheen. The flower buds are like bunches of small, yellow-brown bananas. Hundreds of winged seeds are contained within smooth, woody seed pods that are up to 8 in (20 cm) long and 2 in (5 cm) wide.

Best in bloom
One of the finest flowering trees of the tropics, the African tulip tree is primarily native to Uganda.

PROFILE

- 50 ft (15 m)
- Up to 30 ft (10 m)
- Evergreen
- Opposite simple
- N.W. South America

leaf deep green on upper surface

Smooth leaves
The leaves are ovate to lanceolate, with a slight sheen on the upper surface.

pale green underside of leaf

leaf up to 6 in (15 cm) long

FEATURES

Bark Gray, smooth at first; becomes fluted in maturity

Flowers Fragrant, white to pink, occasionally red; up to ¼ in (5 mm) wide

Cinchona officinalis (Rubiaceae)

QUININE TREE

The discovery of this broadly spreading tree led to one of the most important medical advancements in human history. The bark of the quinine tree and related *Cinchona* species is the only known natural source of quinine, a chemical compound that is an effective treatment for malaria.

This tree is native to the western region of the Amazon rainforest, in Ecuador and Peru, and it was the Quechua Indians of Peru who discovered its medicinal properties. By the early 17th century, the Jesuits had introduced it into Europe.

As the tree matures, the trunk develops deep, long fluting. The leaves, up to 2½ in (6 cm) wide, have smooth margins. The small, fragrant flowers are tubular-shaped and borne in open-stalked terminal panicles. Both the flowers and the panicles are covered in fine, silvery-white hairs. The fruit is a green ovoid-shaped capsule, up to ½ in (1.5 cm) long, which ripens to brown and splits open to reveal numerous winged seeds.

GLOSSARY

ACHENE
A small, dry and hard, one-seeded fruit that does not split open for seed distribution.

AGGREGATE FRUIT
A cluster of many ripened ovaries (fruits) produced from a single flower. Examples are strawberry, raspberry, and blackberry.

ALTERNATE
Describes leaves that are borne singly at each node, in two vertical rows, on either side of an axis.

ANGIOSPERM
A seed-producing plant that encloses its seeds in a fruit and produces flowers. All flowering trees are angiosperms.

ANTHER
In flowering plants, the structure on the stamen that contains pollen.

ARIL
Coat that covers some seeds; often fleshy and brightly colored.

AXIL
The angle formed by a leaf or lateral branch with the stem.

AXILLARY
Borne in an axil, usually referring to flowers.

BARK
The outer layers of the stem of a woody plant, which protect the plant from water loss, cold, infection and mechanical damage.

BASAL ANGIOSPERM
See primitive angiosperm.

BIPINNATE
A compound leaf arrangement in which the leaflets are themselves pinnate.

BRACT
Modified leaf at the base of a flower or flowerhead. A bract may be small and scalelike, or large and petal-like, or it may resemble normal foliage.

BRACTEOLE
Secondary bract sheathing a flower in an inflorescence.

BURR
1. Prickly, spiny, or hooked fruit, seed head, or flowerhead.
2. Woody outgrowth on the trunk of some trees.

BUTTRESS
Trunk base that is fluted or swollen, giving stability to a tree in shallow soil conditions.

CALCAREOUS
Describing soil that contains calcium carbonate and so is alkaline in nature.

CALYX (PL. CALYCES)
Collective name for sepals, joined or separate, which form the outer whorl of the perianth.

CAPSULE
A type of dry fruit that splits open to disperse ripe seed—sometimes called a pod.

CARPEL
Female flower part consisting of a style, a stigma, and an ovary.

CATKIN
A type of inflorescence, usually pendulous, in which scalelike bracts and tiny, often petalless flowers are arranged in a spike.

CLADE
A group of species that includes all the evolutionary descendants from a single common ancestor.

CLONE
A genetically identical group of plants derived from one individual by vegetative reproduction.

COLUMNAR
A tree shape that is significantly taller than it is broad, with parallel sides.

COMPOUND
Describes a leaf that is divided into leaflets.

CONE
The seed-bearing or pollen-bearing structure of conifer trees.

CONICAL
Widest at the bottom, tapering toward the top.

COPPICE
An area of woodland where trees are periodically cut back to ground level in order to encourage the growth of new shoots from the stumps.

CORDATE
A leaf that is heart shaped.

COROLLA
1. Collective name for petals.
2. Inner whorl of perianth segments in some monocotyledons.

CORYMB
Broad, flat-topped or domed inflorescence of stalked flowers or flowerheads arising at different levels on alternate sides of an axis.

COTYLEDON
The first food-containing leaf of a plant embryo that is formed inside the seed. Monocotyledon seeds generally have a single cotyledon, dicots (such as eudicotyledons and primitive angiosperms) two, and those of conifers have a variable number.

CULTIVAR
Contraction of "cultivated variety." A cultivated plant that retains distinct characteristics when propagated.

CYME
Branched inflorescence with each axis ending in a flower.

DECIDUOUS
A tree that loses its leaves and remains leafless for some months of the year, usually in winter (temperate zones) or the dry season (tropical zones).

DIGITATE
A palmate compound leaf divided into five leaflets arising from a single basal point.

DIOECIOUS
A plant that has male and female flowers on different plants.

DRUPE
A type of fruit consisting of a soft outer wall and one or more inner hard "stones," each containing a seed. Fruits of cherries, plums and date palms are drupes.

ELLIPTIC
A flat leaf shape that is broadest at the center, tapering toward each end.

EMBRYO
A young plant at a rudimentary stage of development. In seed plants, including seed trees, the embryo is encased in a seed until it grows into a seedling during germination.

ENDOCARP
The inner layer of a fruit immediately around the seed; the endocarp forms the hard "stone" of drupes.

ENDOSPERM
Term given to the specialized food store within the fertilized seed of angiosperms (flowering plants) that nourishes the embryo during germination.

ENTIRE
A leaf margin that is without teeth.

EPIPHYTE
A plant that has no roots in the soil and lives above the ground surface, supported by another plant.

EVERGREEN
A tree that bears leaves throughout the year.

EXOCARP
The outermost "skin" of a fruit.

FASCICLE
A tight cluster of leaves or stalked flowers that diverge very little. For instance, the needlelike leaves of many pine trees arise in fascicles.

FOLLICLE
Dry fruit, formed from a single carpel, that splits along one side to release one or more seeds.

FORM
Any variant of a species.

FRUIT
A structure of a plant that develops from the ovary of a flower and contains one or more seeds. Fruit can be simple, like berries, or compound, where the fruits of separate flowers are merged.

GAMETE
Cell or nucleus that participates in sexual fusion.

GENUS
A category in classification consisting of a group of closely related species, and denoted by the first part of the scientific name, e.g. *Pinus* in *Pinus pinea*.

GERMINATION
The changes undergone by a reproductive body, such as a seed, immediately prior to and including the first indications of growth.

GLABROUS
Hairless.

GLAUCOUS
Covered with a bloom, or bluish white or bluish grey.

GLOBOSE
Spherical.

GYMNOSPERM
A seed-producing plant that doesn't enclose its seeds in a fruit. Many gymnosperms carry their seeds on cones.

HABIT
The growth form of a tree, including its size, shape, and orientation.

HERBACEOUS
A non-woody plant that normally lives for more than one season, usually dying back at the end of the growing season and overwintering by means of underground rootstocks.

HERITAGE TREE
A tree that has some important cultural or historical significance.

HERMAPHRODITE
Any organism that has both male and female sex organs; in a tree, where the male and female reproductive organs are borne on the same tree.

HYBRID
Naturally or artificially produced offspring of genetically distinct parents of different species. Hybrids show new characteristics. Hybrid names include the names of both parent species separated by an "x", for example *Tilia* x *euchlora*.

INDICATOR SPECIES
A species that by its presence (or absence) shows the type of environmental conditions that prevail.

INFLORESCENCE
Arrangement of flowers around a single axis (stem).

LANCEOLATE
Leaf that is broadest below the centre, tapering to a narrow tip.

LATERAL
1. Located on or to the side of an axis.
2. Side shoot arising from the stem of a plant.

LAX
Describes an arrangement of structures, such as a cluster of flowers, that is loose instead of compact and crowded.

LEADING SHOOT
The terminal shoot of a main branch.

LEAFLET
Single division of a compound leaf. Also known as a pinna.

LEAF AXIL
The angle between the upper surface of a leaf and the branch from which it is growing. A bud often grows in the leaf axil. Also, where a vein comes off the midrib of a leaf.

LENTICEL
Raised pore on the surface of bark or some fruits that provides access for air to the inner tissues.

LOBE
A protruding part of a leaf or a flower.

MARGIN
The edge of a leaf blade. Leaf margins may be smooth (entire), lobed, or serrated

MESOCARP
The soft, fleshy, middle portion of the wall of a succulent fruit between the skin and the hard layer.

MIDRIB
Primary (usually central) vein running from the stalk to the tip of a leaf or leaflet. Also called midvein.

MONOECIOUS
A plant that has separate male and female parts on the same individual.

MONOTYPIC
A taxonomic group with a single species, such a family that contains only one genus or a genus with only one species.

MUTATION
A random change in the genetic makeup of an organism, in its genes or chromosomes; mutations may be passed on through generations.

NATIVE
A species that occurs naturally in a particular region, as opposed to one that has been introduced.

NATURALIZED
A non-native species that has become established in the wild in a region where it was introduced.

NODE
Point on a stem, sometimes swollen, at which leaves, leaf buds, and shoots arise.

OBLANCEOLATE
Leaf shape that is broadest above the center, tapering to a narrow basal point. Also called inversely lance-shaped (lanceolate).

OBLONG
Describes a leaf shape with two sides of roughly equal length.

OBLONG-ELLIPTIC
Describes a leaf shape that is oblong with ends that are round.

OBOVATE
A leaf shape that is egg-shaped in outline and broadest above the middle.

OPPOSITE
A leaf arrangement in which the leaves are borne in pairs at each node, in the same place but on opposite sides of an axis.

OVARY
The lower, wide vessel-like section of the female part of a flower that contains one or more ovules. After fertilization—and the ovules become seeds—the ovary develops into a fruit.

OVATE
Describes a leaf shape that is broadest below the middle.

OVOID
Describes a solid form that is broadest below the middle; usually refers to a fruit.

OVULE
Structure that contains the egg of a seed plant. Ovules are encased in an ovary in flowering plants, but naked in gymnosperms (conifers, ginkgo, and cycads). After fertilization, the ovule becomes the seed.

PALMATE
A compound leaf that is lobed or divided in a hand-like fashion into leaflets arising from a single basal point.

PANICLE
An elongated, branched flower cluster, with stalked flowers.

PEALIKE
Describes a flower with an erect standard petal, two lateral wing petals, and two lower, keeled petals enclosing the stamens and pistil.

PEDICEL
The stalk supporting a flower, flowerhead, or fruit.

PETAL
One of the parts of the corolla of a flower. It is often brightly colored to attract pollinating animals.

PETIOLE
The stalk of a leaf.

PERIANTH
Collective term for the corolla and calyx.

PERICARP
The tissues that make up fruit derived from an ovary: the pericarp is divided into exocarp, mesocarp, and endocarp.

PHLOEM
Plant tissue consisting of microscopic vessels that transport water and dissolved food (such as sugars) through the plant.

PHOTOSYNTHESIS
A process whereby a plant uses light energy to make food and oxygen.

PINNA (PL. PINNAE)
One of the main divisions of a feather-like compound leaf, such as a fern frond.

PINNATE
A compound leaf arrangement in which the leaflets (pinnae) are arranged either alternately or in opposite pairs on a central axis.

PIONEER SPECIES
A species that is usually the first to colonize inhospitable sites and cleared ground, leading the way for other vegetation. Pioneer species improve soil quality and protect land from erosion.

PISTIL
Female reproductive organ of a flower, composed of one or several fused or separate carpels.

POLLARD
To prune the tops of trees in such a way as to encourage new growth in the form of a dense head of foliage.

POLLEN
Tiny grains produced by seed plants that contain male gametes (sex cells) for fertilizing the female egg.

PRIMITIVE ANGIOSPERM
A group of flowering plants with certain primitive features that diverged away from the main evolutionary line of flowering plants before others. Also called basal angiosperms.

RACEME
Inflorescence of stalked flowers radiating from a single, unbranched axis, with the youngest flowers near the tip.

SAMARA
A type of dry fruit with papery "wings" that can be carried by wind, thus dispersing seed farther from the parent tree. Samaras are either single-winged (as in most maples) or multi-winged.

SEMI-DECIDUOUS OR SEMI-EVERGREEN
A tree that loses its leaves for only a short period during the year or that sheds a proportion of its leaves periodically but is never entirely leafless.

SEED
Developmental stage of a seed plant, consisting of an encapsulated embryo.

SEPAL
One of the parts of the calyx of a flower, usually small and leaf-like, and enveloping the unopened flower bud.

SERRATED
Describes the margin of a leaf that has sharp "teeth" pointing forward, like the edge of a saw.

SIMPLE
Undivided, as in a leaf.

SPECIES (ABBREV. SP.)
A classification category defining a group of similar plants that usually interbreed, e.g. Scots pine (*Pinus sylvestris*) is one species.

SPERM
A male reproductive cell capable of fertilization.

SPIKE
An inflorescence in which stalkless flowers are arranged on an unbranched axis.

SPORE
A single cell containing half the quantity of genetic material of typical body cells. Unlike gametes, spores can divide and grow without being fertilized. Tree ferns are the only trees that reproduce by spores.

STAMEN
Male part of a flower, usually composed of an anther borne on a filament.

STAMINAL TUBE
A structure formed from joined stamens.

STIGMA
Female part of a flower at the tip of the pistil that receives pollen.

STIPULE
Leaflike or bract-like structure borne at the point where a leaf-stalk arises from a stem.

STOMA (PL. STOMATA)
Tiny adjustable pore on the surface of a plant that allows for the exchange of gases for photosynthesis and respiration.

STRANGLER
A plant that initially uses another plant (known as the host) for support, before out-competing the host for light and water. Several members of the *Ficus* genus are stranglers.

STROBILUS
A cone-like reproductive structure, such as the seed-bearing structure of conifer trees.

STYLE
Female part of flower connecting the ovary and the stigma.

SUBULATE
Awl-shaped.

SUBSPECIES (ABBREV. SUBSP.)
A category of classification, below species, defining a group within a species, often isolated geographically but able to interbreed with others of the same species.

SUCKER
1. Shoot that arises below the soil level, usually from the roots rather than from the stem or crown of the plant.
2. Shoot that arises from the stock of a grafted or budded plant.

SYNCARP
An aggregate or multiple fruit that is produced from fused pistils; the individual fruits join together in a mass and grow together to form a single fruit.

TEPAL
The term that is given to the sepals and petals when they are indistinguishable from one another as in the case of primitive angiosperms.

TERMINAL
The position at the end of a stem or shoot.

TRIFOLIATE
Three leaves, or a leaf made up of three separate leaflets.

UMBEL
Flat or round-topped inflorescence in which numerous stalked flowers are borne in a terminal position from a single point.

VARIEGATED
Having more than one color; usually used to describe leaves.

VARIETY (ABBREV. VAR.)
A naturally occurring variant of a species.

VEIN
A vascular bundle (bundle of transport vessels) running through a leaf.

WHORL
An arrangement of several identical structural parts in a circle around the same point.

For instance, petals make up the whorl of the corolla.

WOODY PLANT
A plant that has wood: a type of strengthening tissue made from thick-walled vessels of xylem tissue. For example; trees, shrub, and woody climbers such as ivy.

XYLEM
Plant tissue consisting of microscopic thick-walled vessels that transport water and dissolved minerals through the plant.

LIST OF TREE FAMILIES

The following is a list of all tree families covered in this book along with the genera included within each family and the relevant page numbers. These family names are the most up to date at the time of going to print. However, readers should be aware they may find alternative family names still being used elsewhere.

Anacardiaceae *Mangifera* (p.330), *Rhus* (pp.308–309)

Aquifoliaceae *Ilex* (pp.279–280)

Araucariaceae *Araucaria* (pp.28–p.29), *Wollemia* (p.30)

Arecaceae *Butia* (p.117), *Chamaerops* (p.115), *Cocos* (p.119), *Phoenix* (p.116), *Trachycarpus* (p.114)

Asparagaceae *Cordyline* (p.118), *Dracaena* (p.120)

Betulaceae *Alnus* (pp.180-183), *Betula* (pp.173–179), *Carpinus* (pp.184–185), *Corylus* (p.186), *Ostrya* (p.187)

Bignoniaceae *Catalpa* (pp.305–306), *Jacaranda* (p.333), *Spathodea* (p.334)

Bixaceae *Bixa* (p.323)

Caricaceae *Carica* (p.325)

Cephalotaxaceae *Cephalotaxus* (pp.31–32)

Cercidiphyllaceae *Cercidiphyllum* (p.212)

Clusiaceae *Clusia* (p.317)

Cornaceae *Cornus* (pp.274–276), *Davidia* (p.272), *Nyssa* (pp.209–210)

Cupressaceae *Athrotaxus* (p.60–61), *Calocedrus* (p.39), *Chamaecyparis* (p.40, pp.43–44), *Cryptomeria* (p.42), *Cunninghamia* (p.68), *Cupressus* (pp.51–52), *Fitzroya* (p.45), *Glyptostrobus* (p.70), *Juniperus* (pp.53–55), *Metasequoia* (p.67), *Platycladus* (p.58), *Sequoia* (p.62), *Sequoiadendron* (p.63), *Taxodium* (p.66), *Thuja* (pp.56–57), *Thujopsis* (p.59), x *Cuppressocyparis* (p.41)

Dicksoniaceae *Dicksonia* (p.23)

Ebenaceae *Diospyros* (p.324)

Ericaceae *Arbutus* (pp.213–215)

Eucommiaceae *Eucommia* (p.124)

Eucryphiaceae *Eucryphia* (pp.223–224)

Euphorbiaceae *Hevea* (p.329)

Fabaceae *Acacia* (p.229–230), *Albizia* (p.228), *Bauhinia* (p.326), *Cercis* (p.227), *Cladrastris* (p.267), *Delonix* (p.327), *Gleditsia* (p.266), *Gymnocladus* (p.265), *Laburnum* (pp.225–226), *Robinia* (p.262), *Sophora* (p.263)

Fagaceae *Castanea* (p.161), *Chrysolepis* (p.162), *Fagus* (pp.152–156), *Lithocarpus* (p.163), *Nothofagus* (pp.157–160), *Quercus* (pp.164–172)

Ginkgoaceae *Ginkgo* (p.26)

Hamamelidaceae *Liquidambar* (pp.207–208), *Parrotia* (p.211)

Juglandaceae *Carya* (pp.148–151), *Juglans* (pp.141–143)

Lamiaceae *Tectona* (p.332)

Lauraceae *Cinnamomum* (p.112), *Laurus* (pp.107–108), *Persea* (p.111), *Sassafras* (p.127), *Umbellularia* (p.128)

Lecythidaceae *Bertholletia* (p.322)

Lythraceae *Lagerstroemia* (p.231)

Magnoliaceae *Liriodendron* (p.105–106), *Magnolia* (pp.96–103), *Michelia* (p.104)

Malvaceae *Adansonia* (p.319), *Ceiba* (p.320), *Chorisia* (p.321), *Theobroma* (p.318), *Tilia* (pp.188–192)

Meliaceae *Swietenia* (p.331)

Monimiaceae *Laurelia* (p.109)

Moraceae *Artocarpus* (pp.310–311), *Ficus* (p.146, pp.312–316), *Maclura* (p.147), *Morus* (pp.144–145)

Myrtaceae *Eucalyptus* (pp.268–271), *Luma* (p.126), *Metrosideros* (pp.122–123)

Oleaceae *Fraxinus* (pp.300–304), *Ligustrum* (p.130), *Olea* (p.129), *Phillyrea* (p.131)

Pinaceae *Abies* (pp.71–75), *Cedrus* (pp.46–48), *Larix* (pp.76–78), *Picea* (pp.80–93), *Pseudolarix* (p.79), *Pseudotsuga* (pp.49–50), *Tsuga* (pp.64–65)

Pittosporaceae *Pittosporum* (p.132)

Platanaceae *Platanus* (pp.204–206)

Podocarpaceae *Podocarpus* (p.34), *Prumnopitys* (p.33) *Saxegothaea* (p.35)

Proteaceae *Embothrium* (p.125)

Rhizophoraceae *Rhizophora* (p.328)

Rosaceae *Crataegus* (pp.243–245), *Malus* (pp.246–249), *Mespilus* (p.261), *Prunus* (pp.232–242), *Pyrus* (pp.250–251), *Sorbus* (pp.252–260)

Rubiaceae *Cinchona* (p.335)

Rutaceae *Phellodendron* (p.277), *Tetradium* (p.278)

Salicaceae *Populus* (pp.193–197), *Salix* (pp.198–203)

Sapindaceae *Acer* (pp.287–299), *Aesculus* (pp.281–286), *Koelreuteria* (p.273)

Sciadopityaceae *Sciadopitys* (p.69)

Scrophulariaceae *Paulownia* (p.307)

Simaroubaceae *Ailanthus* (p.264)

Styracaceae *Halesia* (pp.219–220), *Styrax* (pp.216–218)

Taxaceae *Torreya* (p.36), *Taxus* (pp.37–38)

Theaceae *Stewartia* (pp.221–222)

Ulmaceae *Celtis* (pp.137–138), *Ulmus* (pp.133–136), *Zelkova* (pp.139–140)

Winteraceae *Drimys* (p.110)

INDEX

ACKNOWLEDGMENTS

Produced in collaboration with the **Smithsonian Institution** in Washington D.C., USA.

Smithsonian Enterprises:
Kealy Wilson, Product Development Coordinator; Ellen Nanney, Licensing Manager; Brigid Ferraro, Director of Licensing; Carol LeBlanc, Vice President

The publisher would like to thank the following: Derek Harvey for additional text and advice; Jane Parker for the index; Monica Byles for proofreading; Lili Bryant for editorial assistance; Sophia Tampakopoulos, Jacket Design Development Manager; Jonny Burrows for icon design; Jacqui Swann and Fiona Macdonald for design assistance; Sam Priddy, Fiona Macdonald, and Jonny Burrows for their help on the photoshoots; and Jenny Faithfull and Julia Harris-Voss for additional picture research.

Thanks to the following for assistance with photoshoots: Katrina Podlewska at The Forestry Commission National Arboretum Westonbirt, Gloucestershire and the Westonbirt volunteers Mike Westgate, Rosemary Westgate, Val Taylor, Peter Higginbotham, Brian Marsh, Sue Bradley Jones; Matthew Hall at Batsford Arboretum, Gloucestershire; and Andrew Conner at The Royal Botanic Gardens, Kew, London

The publisher would like to thank the following for their kind permission to reproduce their photographs:

(Key: a-above; b-below/bottom; c-center; f-far; l-left; r-right; t-top)

1 Corbis: Fumio Tomita/amanaimages. **3 Getty Images:** Laszlo Podor Photography. **5 Tony Russell:** (bl, br). **6 Dorling Kindersley:** Thomas Marent (tr). **7 Alamy Images:** Martin Hughes-Jones (br). **9 Dorling Kindersley:** Alan Buckingham (cr). **11 Dinesh Valke:** (br). **13 Dorling Kindersley:** Paul Whitfield (tr). **14 Corbis:** John Buckingham (tr). **20-21 Corbis:** Mark Karrass. **23 Dorling Kindersley:** The Royal Botanic Gardens, Kew (cl, bl). **26 Tony Russell:** (cl). **27 Getty Images:** National Geographic/Ralph Lee Hopkins (b). **28 Tony Russell:** (t). **29 Ardea:** John Mason (JLMO) (cl). **30 Tony Russell:** (cl, b, t). **34 Dorling Kindersley:** The Forestry Commission at Westonbirt, The National Arboretum (cl, br); The Royal Botanic Gardens, Kew (t). **Tony Russell:** (bl). **36 Tony Russell:** (bl). **37 Tony Russell:** (cl). **38 Tony Russell:** (t, b). **39 Tony Russell:** (bl). **40 Tony Russell:** (cla). **41 Dorling Kindersley:** Batsford Arboretum (t, cl). **44 Dorling Kindersley:** Batsford Arboretum (t, cl). **45 Tony Russell:** (bl). **46 Tony Russell:** (cl). **47 Tony Russell:** (cl). **48 Tony Russell:** (b). **49 Tony Russell:** (bl, cl, tr). **50 Tony Russell:** (cl). **51 Tony Russell:** (cl). **52 Dorling Kindersley:** (t); Batsford

Arboretum (cl). **53 Tony Russell:** (cl). **54 Tony Russell:** (tl). **55 Dorling Kindersley:** The Forestry Commission at Westonbirt, The National Arboretum (t). **56 Dorling Kindersley:** Batsford Arboretum (t, cl). **59 Tony Russell:** (bl). **60 Tony Russell:** (bl, tr). **61 Tony Russell:** (cl). **62 Tony Russell:** (cl). **63 Dorling Kindersley:** Paul Whitfield © Rough Guides (cl). **64 Tony Russell:** (cl). **65 Tony Russell:** (cl, bl). **66 Tony Russell:** (bl). **68 Tony Russell:** (tl, cl). **69 Dorling Kindersley:** Batsford Arboretum (t, cl). **70 Dorling Kindersley:** The Forestry Commission at Westonbirt, The National Arboretum (cl). **72 Corbis:** Harpur Garden Library (t). **73 Dorling Kindersley:** The Royal Botanic Gardens, Kew (cl). **74 Tony Russell:** (cl). **78 Tony Russell:** (cla). **79 Dorling Kindersley:** The Forestry Commission at Westonbirt, The National Arboretum (b). **80 Tony Russell:** (cl). **81 Tony Russell:** (cl). **83 Tony Russell:** (b). **84 Tony Russell:** (cl, bl). **85 Tony Russell:** (cl). **86 Tony Russell:** (cl). **88 Getty Images:** Steve Taylor (cl). **89 Tony Russell:** (cl). **91 Alamy Images:** Krystyna Szulecka Photography (t). **Tony Russell:** (bl, cl). **93 Tony Russell:** (bl). **96 FLPA:** Francois Merlet (bl). **99 Tony Russell:** (t, b). **100 Tony Russell:** (t). **102 Dorling Kindersley:** The Forestry Commission at Westonbirt, The National Arboretum (cl); The Royal Botanic Gardens, Kew (t). **103 Alamy Images:** Roger Phillips (t). **Tony Russell:** (b). **104 Tony Russell:** (t). **105 Alamy Images:**

Science Photo Library (br). **Tony Russell:** (cl). **106 Alamy Images:** FLPA (bl). **Dorling Kindersley:** (t); The Royal Botanic Gardens, Kew (cl). **108 Tony Russell:** (t). **109 Dorling Kindersley:** The Forestry Commission at Westonbirt, The National Arboretum (t, b). **111 Dorling Kindersley:** The Royal Botanic Gardens, Kew (cl). **114 Tony Russell:** (br). **115 Dorling Kindersley:** The Royal Botanic Gardens, Kew (cl). **116 © Plantdatabase.co.uk:** (t). **Tony Russell:** (cl). **117 Getty Images:** Garden Picture Library/Mark Bolton (t). **118 Dorling Kindersley:** The Royal Botanic Gardens, Kew (t, cl, bl). **120 Alamy Images:** Blickwinkel (cl). **122 Tony Russell:** (t). **123 Dorling Kindersley:** The Royal Botanic Gardens, Kew (t, cl). **124 Tony Russell:** (cl). **126 Getty Images:** Matt Anker (bl). **131 Dorling Kindersley:** The Forestry Commission at Westonbirt, The National Arboretum (cl). **133 Alamy Images:** Cath Evans (bl); tbkmedia.de (tr). **134 Dorling Kindersley:** The Royal Botanic Gardens, Kew (t, cl). **135 Alamy Images:** Frank Blackburn (t). **136 Alamy Images:** Steve Taylor ARPS (t). **Dorling Kindersley:** The Forestry Commission at Westonbirt, The National Arboretum (cl, bl). **137 Dorling Kindersley:** The Forestry Commission at Westonbirt, The National Arboretum (t, cl). **Peter Greenwood:** (b). **138 Dan Mullen:** (t). **139 Dorling Kindersley:** The Royal Botanic Gardens, Kew (t, cl). **140 Dorling Kindersley:** Batsford Arboretum (t, cl). **141 Tony Russell:** (cl). **144 Alamy Images:** Blickwinkel (bl); mauritius images GmbH (cl). **148 Alamy Images:** Blickwinkel (bl). **149 Dorling Kindersley:** The Forestry Commission at Westonbirt, The National Arboretum (cl, t); The Royal Botanic Gardens, Kew (bl). **150 Dorling Kindersley:** The Forestry Commission at Westonbirt, The National Arboretum (cl, bl); The Royal Botanic Gardens, Kew (t, clb). **151 Alamy Images:** Blickwinkel (bl). **153 Dorling Kindersley:** The Forestry Commission at Westonbirt, The National Arboretum (t). **154 Dorling Kindersley:** Batsford Arboretum (cl); The Forestry Commission at Westonbirt, The National Arboretum (bl). **155 naturepl.com:** Nature Production (t). **156 Dorling Kindersley:** The Forestry Commission at Westonbirt, The National Arboretum (t). **Tony Russell:** (bl). **157 Dorling Kindersley:** The Forestry Commission at Westonbirt, The National Arboretum (t, cl). **159 Dorling Kindersley:** The Royal Botanic Gardens, Kew (t). **160 Dorling Kindersley:** The Forestry Commission at Westonbirt, The National Arboretum (t, cl). **163 Steven J. Baskauf:** (t). **Tony Russell:** (bl). **164 Dorling Kindersley:** Batsford Arboretum (t, cl). **165 Tony Russell:** (cl). **166 Dorling Kindersley:** Batsford Arboretum (t, bl). **167 Tony Russell:** (cl). **170 Dorling Kindersley:** Batsford Arboretum (t). **The Royal Botanic Gardens, Kew** (cl). **171 Alamy Images:** CuboImages srl (bl). **Dorling Kindersley:** The Forestry Commission at Westonbirt, The National Arboretum (t,

cl). **172 Alamy Images:** Shapencolour (bl). **Dorling Kindersley:** The Forestry Commission at Westonbirt, The National Arboretum (t, cl). **183 Alamy Images:** Florapix (t). **Corbis:** Martin B. Withers/ Frank Lane Picture Agency (cl). **185 Dorling Kindersley:** The Forestry Commission at Westonbirt, The National Arboretum (cl). **186 Dorling Kindersley:** The Forestry Commission at Westonbirt, The National Arboretum (t, cl, bl). **187 Dorling Kindersley:** The Forestry Commission at Westonbirt, The National Arboretum (t, cl). **194 Dorling Kindersley:** The Forestry Commission at Westonbirt, The National Arboretum (cl). **Fotolia:** alessandrozocc (bl). **195 Photolibrary:** (t). **196 Dorling Kindersley:** The Forestry Commission at Westonbirt, The National Arboretum (cl, bl, t). **197 Dorling Kindersley:** The Forestry Commission at Westonbirt, The National Arboretum (cl, t). **199 Tony Russell:** (bl). **200 Dorling Kindersley:** The Forestry Commission at Westonbirt, The National Arboretum (cl). **201 Dorling Kindersley:** The Forestry Commission at Westonbirt, The National Arboretum (bl, cl). **202 Dorling Kindersley:** The Forestry Commission at Westonbirt, The National Arboretum (cl). **Tony Russell:** (bl). **203 Tony Russell:** (bl). **204 Tony Russell:** (bl). **205 Tony Russell:** (cl). **207 Dorling Kindersley:** Batsford Arboretum (cl). **208 Dorling Kindersley:** The Forestry Commission at Westonbirt, The National Arboretum (t, cl, bl). **209 Dorling Kindersley:** The Forestry Commission at Westonbirt, The National Arboretum (cl). **210 Dorling Kindersley:** The Forestry Commission at Westonbirt, The National Arboretum (t, cl). **211 Dorling Kindersley:** Batsford Arboretum (t). **Tony Russell:** (b). **212 Dorling Kindersley:** (t); Batsford Arboretum (cl). **215 Fotolia:** Chungking (b). **218 Tony Russell:** (bl). **220 Dorling Kindersley:** Batsford Arboretum (t, cl). **Tony Russell:** (bl). **222 Dorling Kindersley:** The Forestry Commission at Westonbirt, The National Arboretum (t, cl); The Royal Botanic Gardens, Kew (bl). **225 Tony Russell:** (bl). **232 Tony Russell:** (cl, tr). **233 Dorling Kindersley:** The Forestry Commission at Westonbirt, The National Arboretum (t, cl). **234 Tony Russell:** (bl). **239 Alamy Images:** imagebroker (t). **240 Alamy Images:** WILDLIFE GmbH (bl). **242 Tony Russell:** (cl). **243 Tony Russell:** (cl). **245 Dorling Kindersley:** The Forestry Commission at Westonbirt, The National Arboretum (cl); The Royal Botanic Gardens, Kew (t). **246 Dorling Kindersley:** The Royal Botanic Gardens, Kew (cl). **Tony Russell:** (t). **249 Tony Russell:** (cl). **250 Getty Images:** Thomas Marent (bl). **253 Getty Images:** Bob Gibbons/ Oxford Scientific (t). **254 Alamy Images:** Blickwinkel (bl). **258 Photolibrary:** Anne Green - Armytage (bl). **259 Tony Russell:** (bl). **262 Tony Russell:** (bl). **263 Dorling Kindersley:** Batsford Arboretum (t). **Tony Russell:** (bl). **264 Dorling Kindersley:** Batsford Arboretum (t, cl). **265 Wikipedia:** hardyplants (bl). **266 Tony

Russell:** (cl). **267 Dorling Kindersley:** The Forestry Commission at Westonbirt, The National Arboretum (cl, t). **269 © Plantdatabase.co.uk:** (cl). **271 James Wood:** (t). **272 Dorling Kindersley:** Batsford Arboretum (cl, bl). **274 Tony Russell:** (bl). **276 Tony Russell:** (bl). **280 Dorling Kindersley:** The Forestry Commission at Westonbirt, The National Arboretum (cl, t). **Tony Russell:** (cl). **283 Alamy Images:** Blickwinkel (bl). **Tony Russell:** (t). **285 Alamy Images:** Blickwinkel (bl). **287 Tony Russell:** (cl). **289 Tony Russell:** (cl). **295 Tony Russell:** (bl, t). **296 Dorling Kindersley:** Batsford Arboretum (t, cl). **297 Dorling Kindersley:** Batsford Arboretum (cl, t). **298 Dorling Kindersley:** The Forestry Commission at Westonbirt, The National Arboretum (t). **302 Dorling Kindersley:** The Forestry Commission at Westonbirt, The National Arboretum (t, cl). **304 Dorling Kindersley:** The Forestry Commission at Westonbirt, The National Arboretum (t, cl, clb). **305 Dorling Kindersley:** Batsford Arboretum (cl). **306 Dorling Kindersley:** Batsford Arboretum (cl); The Forestry Commission at Westonbirt, The National Arboretum (bl). **308 Dorling Kindersley:** The Forestry Commission at Westonbirt, The National Arboretum (t, cl). **309 Tony Russell:** (t). **310 Dorling Kindersley:** The Royal Botanic Gardens, Kew (cl). **311 Dorling Kindersley:** The Royal Botanic Gardens, Kew (t, cl). **312 Dorling Kindersley:** The Royal Botanic Gardens, Kew (cl). **313 Dorling Kindersley:** The Royal Botanic Gardens, Kew (cl). **314 Dinesh Valke:** (t). **315 Tony Russell:** (t, b). **316 Dorling Kindersley:** The Royal Botanic Gardens, Kew (t, cl). **317 Eric Hunt:** (t). **318 Dorling Kindersley:** The Royal Botanic Gardens, Kew (t, cl, bl). **319 Alamy Images:** Blickwinkel (bl). **R McRobbie:** (t). **320 Tony Russell:** (cl). **321 Tony Russell:** (bl). **322 FLPA:** Silvestre Silva (t). **Tony Russell:** (clb). **323 Tony Russell:** (cl). **326 Tony Russell:** (t). **327 Alamy Images:** Arco Images GmbH (bl). **Fotolia:** Christian Musat (t). **328 Alamy Images:** Leonid Serebrennikov (t). **FLPA:** Konrad Wothe/ Minden Pictures (t). **Photolibrary:** (t). **329 Tony Russell:** (t). **331 Alamy Images:** Inga Spence (t). **332 Alamy Images:** Sajith Sivasankaran (t). **Photolibrary:** Ron Giling (cl). **333 Tony Russell:** (t). **334 Dorling Kindersley:** The Royal Botanic Gardens, Kew (cl, bl). **Tony Russell:** (t). **335 Alamy Images:** Wildlife GmbH (t). **Tony Russell:** (cl)

Jacket images: *Front:* **Corbis:** Koichi Hasegawa/amanaimages (b); *Spine:* **Corbis:** Koichi Hasegawa/ amanaimages (t)

All other images © Dorling Kindersley
For further information see:
www.dkimages.com